# A HERITAGE FOR
# THE FUTURE

## Queensland
## Department of Transport

Organisational Services Division

## BRUCE PETERS
CP ENG, MIE(AUST)
Principal Manager
Operational Review

Transport House
PO Box 673
FORTITUDE VALLEY Q 4006

Telephone (07) 253 4657
After Hours (07) 376 1725
Facsimile (07) 253 4905

# KAKADU

## A HERITAGE FOR
## THE FUTURE

PHOTOGRAPHY
### LEO MEIER

TEXT
### SIMON BALDERSTONE

WELDONS

Published by Weldons Pty Ltd
43 Victoria Street
McMahons Point, NSW 2060, Australia

Project Coordinators: Sheena Coupe, Jane Fraser
Editor: Carson Creagh
Map: June Jeffries
Index: Diane Regtop
Design: Warren Penney

© Weldons Pty Ltd, 1987

National Library of Australia
Cataloguing-in-publication data

Meier, Leo, 1951-
Kakadu

Bibliography.
Includes index.
ISBN 0 947116 49 4.

1. Kakadu National Park (N.T.) – Description –
Views. 2. Kakadu National Park (N.T.) – Description.
I. Balderstone, Simon. II. Title.
994. 29'5

Typeset by Amazing Faces, Sydney
Printed by Kyodo-Shing Loong Printing Industries Pte
Ltd, Singapore
Printed in Singapore

**A KEVIN WELDON PRODUCTION**

# FOREWORD

Kakadu – national park, Aboriginal land and world heritage area – is one of Australia's most significant wilderness areas. It has also become one of the continent's most controversial sites, with recent tussles over the relative rights of the Aboriginal traditional owners and of mining companies determined to exploit the undoubted riches beneath its soil. To a public becoming increasingly aware of Kakadu, it is the place where crocodiles still attack, where tourists gather to marvel at the wonders of the landscape and to experience something of a previously unfamiliar Australia.

Kakadu is indeed a place of exceptional beauty, where a variety of extraordinary landforms change their character according to seasonal patterns, where an ancient cultural heritage is displayed through artworks of international significance, where animals, birds and natural vegetation flourish. The living culture of the Aboriginal inhabitants of Kakadu dates back for 50 000 years, providing a link with the most ancient inhabitants of the continent. This heritage has been recognised by the handing over of Stage 1 and part of Stage 2 of Kakadu National Park to its Aboriginal owners. The challenge yet to be taken up is the inclusion of all the South Alligator catchment into a single national park.

Kakadu extends from the mangrove-fringed coastal tidal plains through the floodplains and lowland hills to the sandstone cliffs of the Arnhem Land escarpment. It is an area of remarkable diversity of landform. Perhaps nowhere else does change come so obviously. In the Dry the land seems to gasp for water, the plains teem with wildlife but many plants wither and die, leaving only their roots and seeds ready to begin life anew. Then with the coming of the monsoonal storms of the Wet season, the land bursts into life, plants bud and grow with extraordinary speed, animals breed, and land itself takes on different hues. The cycle has been repeated for millennia as Kakadu responds to the natural cycles of heat and rain, Wet and Dry, life and death.

Cockatoos congregate on a gnarled eucalypt.

# CONTENTS

# AN ANCIENT LAND

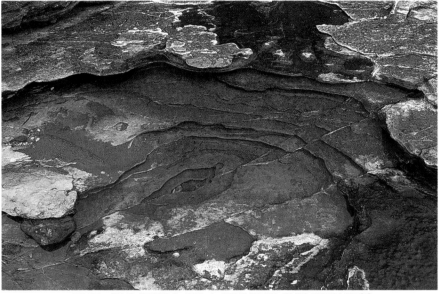

## THE CREATION OF A UNIQUE WILDERNESS

Above: Sun and rain continue to shape
Kakadu's landscapes.

Left: Pandanus frames the dense forests of
the floodplains.

n 1912, when Professor Baldwin Spencer joined buffalo hunter Paddy Cahill at Oenpelli on the East Alligator River, he was entering a region that has since been hailed as one of the world's most valuable and important wilderness areas. Introduced by Cahill to the local Aborigines, Spencer wrote that, 'The Kakadu is one of a group, or nation, of tribes inhabiting an unknown extent of country, including that drained by the Alligator Rivers, the Cobourg Peninsula and the coastal district, at all events as far west as Finke Bay. Its eastern extension is not known. For this nation I propose the name of Kakadu, after that of the tribe of which we know most.'

The *Kakadu* were in fact *Gagudju* or *Ka:kudju*, members of a specific language group and part of a population that had probably numbered more than 2000 when Ludwig Leichhardt traversed the region in 1845, marvelling at the Aborigines, the wildlife and the diverse topography.

They are now only six Gagudju speakers left, but Spencer's naming has endured and is now institutionalised as the Kakadu National Park. The park dominates the region, which still includes the greatest legacy of the Gagudju: five thousand sites of ancient and incomparable rock art, designated in 1979 as 'one of the richest concentrations of superb rock art in the world ... It is heritage without parallel, and like all unique treasures is beyond evaluation.'

This is the main component of the Kakadu's cultural heritage: a heritage older than any in Australia. The very first Australians lived in the area after their arrival more than 40 000 years ago and depicted those millennia in vivid, detailed paintings. Kakadu is a window through which we can all look at the human and natural history of the land.

Moreover, it contains a diversity of environments, plants and animals not found elsewhere. One hundred and nineteen years after Leichhardt wrote glowingly of the 'magnificently grassed' South Alligator River region, a 'beautiful valley ... the promised land', and that, 'the cackling of geese, the quacking of ducks, the sonorous note of the native companion, and the noises of black and white cockatoos and a great variety of other birds, gave to the country, both night and day, an extraordinary appearance of animation,' the biological significance of the area was recognised when the Woolwonga Aboriginal Reserve was proclaimed a wildlife sanctuary.

Towering stormclouds herald the annual transformation of
Kakadu's landscapes in an explosion of life and growth.

Governments were advised that there was 'no time to waste' in setting aside an area for what could be 'one of the best national parks in the world'.

Similar recommendations were forthcoming in 1969 and 1970: 'The National Park Proposal in the Northern Territory is of national importance to all Australians in the preservation of an area of outstanding natural resources, comparable with any of the great parks of the world,' said a Department of the Interior planning team report in 1970. 'The natural features are unique, wildlife is prolific. A major attraction of the National Park is the grandeur of the rock formations of the escarpment and the magnificent panoramas from its vantage points'.

The Alligator Rivers Wildlife Sanctuary, covering 3290 square kilometres, was established in 1972, but it was not until 1978 that firm arrangements began to be made for an extensive National Park. The Commonwealth acquired for National Park purposes land that became Stages One and Two. The Gimbat and Goodparla pastoral leases to the south, which included the environmentally influential upper catchment of the South Alligator River, were also acquired.

Stage One of Kakadu National Park was proclaimed on 5 April 1979. It totalled 6144 square kilometres and took in a central section of the sandstone escarpment that snakes its way from north to south through the region. Jim Jim and Nourlangie Creeks, running from the escarpment, cross the park's lowlands and plains to the South Alligator River on the western edge of the park. At its eastern edge Jim Jim Falls and Twin Falls plummet to the foot of the escarpment, feeding both the rivers and the aesthetic appetites of visitors.

The extensive art galleries at Nourlangie Rock (perilously close to the proposed Koongarra uranium mine) and at Ubirr, on a northeastern finger of the Stage One area, are also in the park.

For the Aborigines, these paintings are an integral link with a land rich in cultural and spiritual legacies. The Kakadu region contains thousands of galleries, sacred sites and features that to this day are of deep, personal importance and impressive aesthetic beauty.

Much of the landscape, 'natural' to Europeans, is the product of the actions, movements, lives and deaths of the ancestors who fashioned the land, gave it all living things, then took on a particular form themselves as well as the Aboriginal hunters and 'fire-stick' farmers of Kakadu.

Right: Smoke from a Dry season fire veils the escarpment.
Overleaf: Twin Falls carves a path to the plains below.

The Aboriginal relationship with the land is a manifestation of this reverence. The land and its people are in partnership. There is no 'master-servant' relationship; to lose the land is to lose part of yourself and you must adapt to the land, not force it to adapt to your needs.

In 1978 the traditional Aboriginal owners, having won rights to much of the Alligator River region, leased land back to the Australian National Parks and Wildlife Service to be managed as a National Park. The arrangement included the controversial decision by the Federal Government to allow uranium mining at Ranger, necessitating Aboriginal agreement and the establishment of substantial infrastructure, including a town, on the edge of Stage One of the park.

The Ranger Uranium Environmental Inquiry had recommended to the Federal Government in 1977 that mining proceed under strict environmental guidelines, that land rights be granted, and that the establishment of a National Park was central to any decisions about the region.

Stage One also included a separate area to the north, on the coast of Van Diemen Gulf, where Magela Creek, having passed the Ranger uranium mine, joins the East Alligator River and they flow to the sea.

The World Heritage Committee of the United Nations placed Kakadu National Park on the World Heritage List in October 1981. The 6144 square kilometres of wetlands, plains, escarpment and unique wildlife and Aboriginal art had become the first part of Australia to be so honoured.

The minutes of the Committee meeting contained an important addition: 'The Committee noted that the Australian Government intended to proclaim additional areas in the Alligator River region as part of Kakadu National Park and recommended that such areas be included in the site inscribed in the World Heritage List and that in the region the environmental protection measures specified in the relevant legislation continue to be enforced.'

The World Heritage List is not an easy club to join. The United Nations Educational, Scientific and Cultural Organisation (UNESCO) obtains independent confirmation of the outstanding heritage value of the proposed area, and two international panels of scientists and other experts assess its cultural worth and natural significance. Both were in full support of the inclusion of Kakadu. In 1984 a further 6929 square kilometres became Stage Two of the National Park.

Left: The rushing waters of a swollen Wet season creek.
Overleaf: The escarpment gives way to a flooded swampland.

This stage included Field and Barron Islands and the vast and biologically important wetlands, plains and woodlands around the West and South Alligator and Wildman Rivers and almost the entire region between the South and East Alligator Rivers. It filled the gap between the coast and Stage One, except for the significant excision of the Ranger and (still proposed) Jabiluka uranium mine sites. The Koongarra uranium mine site about twenty kilometres south of Ranger is also excised from the Park, as are the leases of the Border Store on the way to Oenpelli and the pleasant tourist facilities at Cooinda, on Yellow Water lagoon, managed by the Gagudju Association.

The town of Jabiru, near its *raison d'être*, the Ranger mine, is within the Park boundary but is managed by an independent authority.

The two stages of Kakadu were proclaimed as an integrated park on 20 December 1985, combining into a single management entity an area some 200 kilometres long from the coast to the southern hills and 100 kilometres wide from western Arnhem Land to the Wildman River region.

However, the area designated Stage Two of Kakadu National Park is still not on the World Heritage List, despite the World Heritage Committee's comments in October 1981. Court action and political pressure, including mining companies' film of Stage Two purporting to show country made 'second-rate' by buffalo damage, succeeded in halting the nomination of the second stage.

While Stage Two has not been granted recognition and protection under World Heritage listing, another 6000 square kilometre section to the south has not even achieved its proposed status at Stage Three.

The Ranger Uranium Environmental Inquiry's report mentioned the desirability of including a complete river catchment in the Park to protect water quality and recommended the inclusion of the Gimbat and Goodparla pastoral leases to the south, thus permitting the inclusion of most of the headwaters of the South Alligator River.

If this catchment were to be included, erosion and other damage that have direct effects on the South Alligator system could be significantly decreased and the lowlands and their wildlife could be protected fully.

The wetlands are remarkable. More than a third of all the birds recorded in Australia have been observed in Kakadu. Thousands of waterbirds, including

Life clings tenaciously to the rock, seeking out every drop of
moisture to survive the long months of the Dry season.

jabiru storks, brolgas, magpie geese, herons and jacana feed and breed on the floodplains and coastal wetlands during the Dry season. As the Wet approaches many species leave the region, their place being taken by the wading birds that return from Asia to the grasslands. Eighty or so non-breeding species use the park as a temporary refuge during their annual migrations between the Northern Hemisphere and Australia.

The open woodlands and forested areas of Kakadu also support a variety of birds and the estuarine and coastal mangroves are inhabited by yet another distinct bird fauna.

Several species of birds are known only from Kakadu, and the park's wealth of birdlife has prompted the United Nations to place the wetlands on its list of "Wetlands of International Importance especially as Waterfowl Habitat'.

The movement of bird population is dictated by Kakadu's wet-dry tropical monsoonal climate. The area has an average annual rainfall of around 1400 centimetres, and monthly average temperatures vary from 25° to 30°C.

Almost ninety per cent of the rain falls between November and March; the over-simplified modern categorisation of the climate dissects the year into the Dry season – April to October – and the Wet – November to March.

The Aborigines, with a broader and more detailed perspective on the climate, divide the year into many seasons, each with its chain of natural events that fit into an overall pattern of informed exploitation of the environment.

The climate, however it is interpreted, also has a profound effect on the vegetation and landscape. The park encompasses tidal flats and coastal swamps, grassland and sedgeland blacksoil floodplains, open and closed forest and woodland, rainforests and the imposing rock escarpment and plateau that themselves host a variety of communities, from sparse spinifex to unique canopy rainforest.

There are at least a thousand (and possibly as many as 1200) species of plants in the park. Each survey and research project increases the number of wildlife species recorded, but it is now established that there are at least 10 000 insect, fifty-five fish, twenty-five frog, seventy-five reptile, 275 bird and fifty-one mammal species in the region.

Although the traditional relationships, established over millennia, between the Aborigines and the environment have been greatly dissipated by outside

Many of Kakadu's birds find shelter, food and secure nesting sites in the park's waterlily-covered billabongs.

influences they survive in the legends, the paintings, and in the lives of the surviving traditional owners. There are now between 200 and 250 Aborigines living in Kakadu National Park, at such places as Cannon Hill, Jabiru East, Deaf Adder, Nourlangie Rocks and Spring Peak.

The art of Kakadu is one of many images: lily-edged billabongs, spectacular waterfalls, broad stretches of floodplain covered in colourful birdlife and the irridescent foliage of paperbark stands as the rising waters of the Wet play around their roots.

There are less positive images – of buffalo charging through wetlands, and of erosion caused by the disturbance of natural vegetation. The pressures on Kakadu from exotic fauna and human influence are great; one of the paradoxes of such a beautiful place attracting so many thousands of human beings who wish to experience its beauty and its natural and cultural heritage.

In the words of environmental philosopher Val Plumwood, Kakadu is one of the few places that 'distil the need to come to terms with the dark side of the natural world, its capacity to destroy as well as nurture'.

We must respect and take into account the land's power, not seeing the natural world as romantically benign as is possible in southern Australia. The power can be felt very strongly in the land itself, especially in the stone outlier and escarpment country that makes up a fifth of the park and in the presence of creatures such as saltwater crocodiles.

'Aboriginal people know both spirits, the nurturing and destructive,' says Val Plumwood. 'We don't: we downplay both its nurturing and protective capacity and its destructive power so as to aggrandise ourselves and our own power, and try to see it as simply neutral. It is very hard to do that in Kakadu'.

The survival of Kakadu depends on our recognition that the ultimate power lies with nature, and that the wildlife of Kakadu, its landscapes and cultural heritage, are all in tune with such an understanding.

Right: The Kakadu region is rich in spiritual value to its Aboriginal owners.

Overleaf: Wet season floodwaters cover the coastal plains.

The jumbled rocks of the escarpment's outliers offer these sharp-
tipped grasses protection from wallabies and rodents.

Above: The twisted skeleton of a relict rainforest tree beside the almost dry Nourlangie Creek symbolises Kakadu's diversity of life forms.

Overleaf: The rhythm of life slows during the dry winter months.

Above: Each year's rains corrode a little more of the escarpment.

Right: A gnarled paperbark is framed by Wet season grasses.

The colours of the landscape intensify as cyclonic clouds gather over the escarpment, bringing promise of more rain to Kakadu.

Above: A monsoon thunderstorm sweeps over the escarpment.
Overleaf: White trunks of paperbarks etched against floodwaters.

Above: Delicate water plants are easily destroyed by pollution or damage
from introduced animals.

Right: The park offers a rich variety of landscapes and lifeforms.

Overleaf: Protection of Kakadu's southern section is vital to the
preservation of every environment in this unique region.

# A CANVAS OF LANDSCAPES

FROM COASTAL SWAMPS TO RUGGED PLATEAUX

Above: Sere grasses await the monsoon rains
that bring the escarpment to life.

Left: Paperbarks mark the courses of Kakadu's
many creeks and rivers.

High on the Arnhem Land Plateau, above the imposing escarpments on its western edge, there is a series of rippled rock surfaces. They look like the dark sands of a beach at low tide, and that is exactly what they are. The sandstone was deposited in a shallow sea almost 2000 million years ago, and tells one of the oldest stories of the Kakadu Region.

The sandstone, now superimposed over even older volcanic metamorphic and granitic rock, was metamorphosed by heat and pressure and consolidated. Between 1800 and 1400 million years ago these formations were covered by the enormous Kombolgie Formation, which now forms the basis of the Arnhem Land plateau, revealing itself as the escarpment and in outlier formations such as Mount Brockman rising from the lower river plains. The Kombolgie Formation is now predominantly sandstone, with some volcanic rocks. Erosion has removed the formation from several areas in the north and west, but it still covers a large portion of the east and southeast Alligator Rivers system.

About 100 million years ago, Mesozoic sandstone and siltstone (known as the Mullaman Beds) covered large sections of the region, but they too have been eroded in most areas to expose the underlying formations. These beds persist in the extreme east and parts of the north of the region.

In the past sixty million years much of the area has been covered by sand, and alluvium and soils have developed. Exposed formations and rock surfaces have been subjected to continuous modification by weathering, leaching, erosion, transport and deposition of materials to produce new land surfaces.

Erosion is continuing, together with soil formation and deposit along the Alligator Rivers System. Erosion on the Arnhem Land plateau is producing large quantities of sand and coastal and river sediments are still accumulating as water-borne material is deposited after being carried from the hinterland.

More than 2000 million years after the first rock of the region was formed, the mineralisation of the Kakadu region is once more of significance to Australia. The widespread and rich deposits of uranium, gold, palladium and other elements are being sought by miners. Several mining exploration and extraction projects have already taken place in the area.

One climatic index indicates that around thirty to fifty thousand years ago the climate of the Alligator River was similar to what it is today, with a considerably

Ubirr, an ancient home for Aboriginal people.

cooler climate during the coldest period of the most recent ice age, about 18 000 years ago. At that time much of the Kakadu area is thought to have been semi-arid woodland. At the end of that period ice melt lifted the sea level to about what it is today and made tidal estuaries of the lower sections of the Alligator Rivers. The formation of mud banks began a transformation of the surrounding environment, separating swamps and flats from the estuaries.

The sea proved an efficient courier, delivering sand into these bars, and the ridges thus formed led to another important stage of land development: the birth of lagoon systems.

The coastal lowlands of the East Alligator River were exposed by the retreating sandstone Arnhem Land plateau and broad valleys were excised in the ancient basement rocks by streams. In places, isolated outliers such as Mount Brockman rose like islands from the lowland ridges. The broad valleys subsequently filled with mud, becoming swampy plains.

Some of the land formation developments are relatively recent. Fewer than 1500 years ago for example, the area around Yellow Water Billabong, now the site of the Cooinda Hotel complex, was a system of saline mud flats. With a warmer climate, these flats have been replaced by freshwater swamps and billabongs.

## THE LANDSCAPE TODAY

Kadadu's topography and environment have been sculpted into a mélange of raw, sometimes dangerous and always beautiful ecosystems.

Human experiences of Kakadu are reminders of the power of nature and of the futility of our attempts to adapt or exploit the environment beyond its ability to change. The animals, plants and inanimate elements of the environment are fitted into a relationship that ignores the boundaries drawn on maps.

The Aborigines of Kakadu also fitted into that order, adapting to the environment rather than attempting to master it. Now they are removed from parts of that relationship, but their sacred sites and paintings are an integral part of the heritage of Kakadu.

The strategic importance of protecting entire ecosystems lies in the preservation of every element, and the apparently unconnected landscapes of Kakadu are in reality inextricably bound together: to hurt one is to wound the others. For every exotic addition there is a withdrawal, a debit from the system.

Right: The East Alligator River passes through the escarpment.
Overleaf: Sunset on a backwater of the upper reaches of the South Alligator River.

The landscapes of the park can be divided differently according to one's criteria, but most agree that there are distinct sections to the region, each adding to its wilderness and aesthetic qualities – qualities whose significance can be determined by describing the variety of land surfaces and topographical features. There are tidal flats and coastal swamps, floodplains, lowland hills, the escarpment and its outliers, the plateau complex and the southern hills and basins, yet to gain due recognition for their importance.

A detailed study of the region in 1977 determined five types of topography, but did not allocate to the escarpment and the outliers a separate identity. To emphasise the importance of this spectacular and influential feature, it has been added to the list of primary landscape sectors.

Despite its ancient characteristics, the landscape of Kakadu is still changing. The escarpment is still retreating, having left behind the outliers. In the rivers sediment is gathering as it has for thousands of years, literally laying the foundations of new habitats: there is a plateau 2000 million years old, and sands delivered by the river system that have barely settled.

COASTAL SWAMPS AND TIDAL FLATS

The raw power of Kakadu is shown in few better ways than the inrush of the tide at the mouth of the South Alligator River. The grey, swirling waters engulf sands and obliterate rivulets created only hours before by the outgoing waters. Mudflats are overtaken by 5-metre tides as the fast flow and heavy turbulence surge over steep river banks and grey cliffs that gain freedom at low tide each day, only to lose it again.

A series of natural levees traps much of this tidal flow, but cyclone damage and feral water buffalo cause breaches of the banks. There is evidence in the succession of ancient beaches and lines of sand and shell that deposition is still taking place. River sediment is deposited between these lines and the coarser sediments form the natural levees that border the meandering rivers.

As the habitat alters from the sea to the estuaries and the back swamps, salinity and flood levels drop. In those areas furthest from the sea the estuarine plains are clothed in low sedge or fleshy, salt-tolerant samphire (*Arthrocnemum*). Coastward, the flats are sparsely vegetated; this is a forbidding, saline environment that experiences alternate flooding and baking sun.

Tidal flats and dense foliage near the confluence of the coast
and the floodplain.

Now mangroves are a central element in the coastal ecosystem. The dark foliage draws a distinct fringing line around the coast, the swamps and the river edges. In places the mangroves are bordered on their landward side by low sand ridges and dunes, with non-eucalypt monsoon forest. The mangrove swamps have saline, muddy soils, and their dense canopy produces a steamy, dank atmosphere that tends, together with the mud, tree roots, saltwater crocodiles, snakes and insects, to place them well down on the list of access priorities for Kakadu's human visitors.

But the mangrove swamps must be seen as proof of the dark and light sides of nature. The 'toothed Mother' has a power that is not always benign, a beauty not always romantic. Mangrove swamps may be uncomfortable, inaccessible and deadly, but they protect the coastline from erosion and their destruction would rob birds, animals and marine life of an essential breeding refuge. These regions are actually a fascinating observatory of wildlife: inconveniences and discomforts aside, the mangrove swamps are a highlight among Kakadu's many natural highlights, and support an array of complex communities of considerable ecological importance to the region.

## THE FLOODPLAINS

Like the Rainbow Snake so revered in the Aboriginal culture of the area, the Alligator Rivers twist smoothly across the floodplains, overflowing their banks seasonally to transform the broad landscape around them.

The floodplain is described technically as 'senile lowland landscape' and adjoins the tidal flats. The plains collect an enormous flow of water from the plateau and escarpment country to the east and from the southern hills that form the extreme of the catchment of the South Alligator River.

During the Wet season this area is inundated by fresh water, which leaves an annual veneer of silt. During the Dry enormous numbers of birds, including many waterbirds, gather in the swamp areas, providing a stirring sight as well as an opportunity to study and photograph dozens of species in one area.

Small sedges less than sixty centimetres high dominate some of the higher, drier areas of the floodplain that are under water for between two and six months. In areas flooded for longer periods, a rich herbaceous swamp vegetation, containing many species of grass and herbs, is dominant.

Above: Early morning serenity on a billabong in the upper reaches of the
South Alligator River.

Overleaf: Heavily swollen, Nourlangie Creek funnels water into the
surrounding floodplain.

Forests of paperbark (*Melaleuca leucadendron*) cover large areas of the coastal plains, frequenting both waterlogged and drier ground. The paperbark is also found along creeks, where it mixes with other evergreen non-eucalypt trees; in permanently wet areas it has superbly complex, arched root systems.

The floodplains are formed of estuarine mud and clay, with sandy loams predominating in the upper reaches. Like the coarser sediments of the tidal levees in the river valleys, this soil breaks away quickly when disturbed.

The floodplains begin where the Nourlangie and Magela Creek systems leave the lowlands. The streams disperse across the plain, the water spreading to flood wide areas across to the East and South Alligator Rivers, eventually to join the estuarine tidal flats. As well as being fed by the creek and river systems and the escarpments to the east, drainage between the main rivers adds to the spectacular volume of water.

Oxbows, billabongs, back swamps and meander curves are all parts of a complex water system in which riverbeds are higher than the outer floodplain margins, and in which overflow from river banks plays a part in the flooding cycle. Magela Creek and similar low tributaries are separated from the rivers by natural levees or higher sections of the plains, and only begin their full flow as the Wet season waters accumulate.

The impermeable clays of the plains give way to lighter soil and, on the better-drained land, savannah and grassland. The wildlife population of the higher sections and rocky outcrops grows with the coming of the Wet and its rising waters. As the water level drops around March and April, the billabongs and river systems re-emerge and the floodplain soils are subject to deep cracking. Permanent freshwater lagoons have formed on the edges of these plains. Fringed by pandanus palms and freshwater mangroves or paperbarks and carpeted with waterlilies, the lagoons are an enduring image of Kakadu, as are the waterbirds that rely on the plains as their Dry season refuges and that gather there in huge numbers every year.

## THE LOWLAND HILLS

As the land rises gradually to the west and the south, level and rolling lowlands emerge. Among these gentle undulations the old courses of rivers and creeks can be traced by chains of billabongs, waterholes and depressions. These

A swathe of floodplain, inundated by the Jim Jim and South Alligator rivers.

superb habitats and the spring-fed, well-vegetated creeks coming from the escarpment provide an abundance of food and moisture for wildlife.

The lowlands rise between the floodplains bordering the East and South Alligator Rivers and extend to the south past Cooinda, at Yellow Water, to the escarpment and the plateau.

Generally free from flooding, the lowland area has an average relief of twenty-five to fifty metres and its gentle undulations are split by rocky hills and ridges topped by deep sands and the residue of weathered sandstone outliers. The lowlands contain tracts of tall open forest dominated by eucalypts. A layer of twenty or so species of palms, small trees and tall shrubs has developed at a height of about four metres below the eucalypt canopy.

Interspersed among the forests are areas of woodland with smaller, more widely spaced trees. The woodlands are also dominated by eucalypts, though in greater variety, and the lower canopy includes a number of perennial grasses.

The vegetation of the lowlands takes many more forms than dry woodland and tall open forests. There are areas of stunted woodland, dwarfed and crooked, on the stony hills in the southwest, and dense non-eucalypt scrub on the sandy or gravelly slopes bordering the coastal plains. Rarely growing higher than eight metres, this scrub is populated primarily by *Melaleuca, Pandanus, Grevillea* and *Acacia*. There are even small sections of dense monsoon rainforest, in wetter areas where fire, which burns much of the lowlands vegetation each year, seldom reaches.

## THE ESCARPMENT AND ITS OUTLIERS

For more than a thousand kilometres a line of exposed sandstone provides a spectacular backdrop – and a fascinating aesthetic, archaeological and environmental focal point – to Kakadu National Park.

This precipitous escarpment contains a bewildering variety of experiences and habitats. The sight of waterfalls plunging over the escarpment to sand- and palm-fringed pools below, Aboriginal paintings in rock shelters used for thousands of years and the ever-changing shades, textures and colours of the rock itself are a constant source of intrigue.

The human significance of the escarpment may lie in its spectacular and imposing beauty for tourists and in its rich religious and artistic relevance for the

Nourlangie Rock, an outlier of the escarpment, gave shelter to
Aborigines and was a base for food-gathering expeditions.

Aborigines of the area, but for its wildlife the escarpment formed by the retreat of the Arnhem Land plateau is a huge natural shelter, distribution boundary and reference point for movement and migration.

The escarpment runs from north to south through the park, rising between fifty and 250 metres above the lowlands and providing water, protection and food for many species just as it has provided them to Aborigines for millennia.

At its southern extent, the scarp rises more than 200 metres in an almost unbroken wall. Further north, the line is lower and much more broken.

The slow retreat of the plateau and variations in weathering have caused the escarpment to vary greatly in form. There are steep cliffs, over which Deaf Adder, Twin and Jim Jim Falls plunge dramatically into pools, and there are stepped promontories of rock with extended talus slopes.

As the setting sun strikes the snaking, weather-beaten line of cliffs and boulders it alters their tones from reds and yellows to rich brown or cream. The shadows stretch on rock pillars and in gorges, dance among caverns, turrets, and jumbled and stacked sandstone blocks and boulders as the spirits so revered by Aborigines come alive.

Erosion along joint and bedding planes and weathering of softer rock strata have created these patterns of relief, and several outliers are actually microcosms of Kakadu as a whole. The Cannon Hill and Ubirr regions with their surrounding floodplains and lowlands are prime examples. The escarpment and its outliers offer unique experiences for the student of Aboriginal art: others seek in them semi-precious stones or rocks that contain ripple marks made by the tides of millions of years ago.

## THE PLATEAU

The spectacular waterfalls that tumble in roaring splendour over the escarpment signal the boundary of one of the most significant landscapes in the region.

The rivers to which these waterfalls provide such spectacular counterpoint were born on and traverse the Arnhem Land plateau, which runs eastward across the Park and beyond. The ancient sedimentary rocks of the plateau were formed about 2000 million years ago, and have weathered and dissected into a trellis pattern of jointed blocks. The rectangular dissections control the drainage pattern, and erosion has created caverns and silent, secret gorges.

Right: Twin Falls in the Wet season.
Overleaf: Ominous clouds herald an approaching storm.

The plateau rises to 520 metres above sea level, but the majority of the mass is about half that height – still well above even the highest points of the lowlands, Mount Cahill (154 metres) and Mount Basedow (200 metres).

Ripple marks in the rocks atop the plateau are evidence of its origins as sandstone was deposited in a shallow sea and consolidated by pressure.

Where water has sliced along fault lines and joints in the sandstone, deep valleys have formed and the plateau is criss-crossed by a network of crevices. Soil collected in the valleys and crevices supports small communities of vegetation in moist, cool microclimates, but about a third of the plateau is bare sandstone; a jumbled mass of blocks that features little or no plant life. A further third, where shallow sandy soils occur, supports heath and spinifex scrub, a diverse community of legumes and myrtle shrubs typical of Australia's arid interior.

In deeper sands there are isolated pockets of sandstone woodland and tall open forest, primarily of eucalypts and many myrtles and leguminous shrubs. Spinifex and wiry grasses provide ground cover in these pockets. Pockets of rainforest are distinctive features of the plateau. This monocultural rainforest, considered a relict of a past climatic age, consists almost entirely of the evergreen myrtle *Allosyncarpia ternata*.

The process of weathering and erosion that has created such an intricate pattern of blocks, pinnacles, slabs and joints on the plateau is continuing as the plateau retreats southeast at a rate of no more than a metre every thousand years.

## THE SOUTHERN HILLS AND BASINS

Edging into Stage One of Kakadu National Park and stretching south and southwest well beyond the park boundaries to cover an area included in the originally proposed Stage Three are the southern hills and basins. Although lacking the spectacular qualities of the escarpment or the teeming wildlife of the floodplains, the area constitutes the upper catchment of the South Alligator River, which dominates much of the ecology and topography of the region.

A series of folded and faulted rocks dictates the flow of the tributary system, which is dominated by ridges of hills up to 300 metres high, lined by cliffs exposing metamorphic rocks. At the southern tip of this area a sandy plateau is covered mainly by medium and low density woodland. The rare *Allosyncarpia* forest also appears on some dissected sandstone sites.

Right: Trees and grasses flourish on the edge of the escarpment.
Overleaf: Nourlangie Rock.

Lowland vegetation, burnt in the Dry season, regenerates rapidly with
the coming of rain. Shoots appear as if from nowhere as the plants
respond with new life to the changed conditions.

Above: Paperbarks in the flooded swamps surrounding the Nourlangie billabong, whose waters have penetrated the surrounding floodplain.

Overleaf: Baroalba Creek, surrounded by monsoonal forest, is a mere trickle in the Dry; here it contains an unusually large body of water.

Above: A subterranean watercourse seeps from the escarpment.

Left: Jim Jim Creek, in flood, forces its way through the rocks.

Overleaf: From this outlier, the wooded lands stretch southeast
toward the main escarpment.

# THE UNTAMED LAND

A SPRAWLING MIXTURE OF PLANT LIFE

Above: Life springs quickly from death in
the tropical climate.

Left: Kakadu's biological richness is of
global significance.

The landscapes of Kakadu have, like almost all of Australia, felt the hand of the Aboriginal hunter and food-gatherer for tens of thousands of years. Aborigines exploited and modified the environment in less obvious (and less obviously destructive) ways than Europeans, but they did change and maintain certain types of habitats to suit their purposes. By burning areas of woodland and forest at specific times they promoted the growth of fire-sensitive plants and aided hunting.

Fires, generally early in the Dry season, were burned in a mosaic pattern and the management of Kakadu National Park has maintained that tradition, which helps maintain diversity of species and protection against wildfire that can destroy fire-sensitive plantlife such as the monsoon forests.

By burning the open forest areas early in the season Aboriginal communities ensured fires were less intense, occurred in smaller and more easily controlled areas and had lower scorch heights.

The plant communities of Kakadu can be divided into seven major categories containing some 1200 species of plants; almost a thousand have been identified, and studies are continuing. So great is the variety of habitats in the region, from saline estuaries to freshwater lagoons and sandstone plateaux, that it could contain nearly half of the Top End's flora even though it comprises only four per cent of the total area of the Northern Territory.

The 12 700-square-kilometre Alligator Rivers region is a mixture of rocky escarpment country, sparse woodland (20 per cent), wetland (13 per cent), medium density forest (10 per cent), open forest (24 per cent), and woodland (33 per cent).

The variety of habitats is matched by and is intricately related to the variety of plant types. There is little similarity between the heath and spinifex of the sandstone plateau and the complex estuarine mangrove communities, but even in generally similar habitats there are greatly differing assemblies of vegetation. In a single region such as the plateau, for example, hard spinifex country can be adjoined by canopy rainforest.

The littoral community of plants along the coastline and the banks of estuaries is complex and ecologically significant. There are twenty-one of the twenty-nine Australian species of mangrove, all supporting abundant wildlife. Large areas of

Ground cover flowers in profusion during the Wet season.

the tidal mud flats are inhabited by salt-tolerant samphire and communities of herbs, sedges, grasses and small shrubs live in the cracked, baked mud that is inundated only by unusually high tides. On the beach ridges species of *Acacia, Eugenia* and members of the pandanus family form pockets of forests.

The mangroves have adapted to their saline environment and constant inundation in several ways. Their roots can withstand salt intake and some species excrete excess salt through their leaves. The lack of available oxygen has been overcome by roots that appear on the trunk above the water or roots that can store oxygen in their soft interiors when exposed at low tide.

The vast floodplains support extensive sedgeland and swamp forest communities, with large stands of paperbark (*Melaleuca leucadendron* and *M. argentea*), beneath which grow annual grasses and herbs.

The sedge and swamp communities differ markedly according to the level and duration of their flooding. Where the floodplains are under water for two to six months, sedgelands dominate. Sedges, grasses and *Sesbania* abound on the lower sections of these areas and species of *Acacia* and the screw palm *Pandanus spiralis* grow on banks of billabongs and streams.

A freshwater species of mangrove, *Barringtonia acutangula*, also grows on the swampier sections of stream and lagoon edges.

Where water occupies the floodplains for six to nine months of the year, herbaceous swamp vegetation dominates. In deeper (sometimes permanent) water lilies and the floating aquatic fern *Azolla pinnata* cover the surface, interspersed with the green and blue 'blooms' of algae.

Large patches of spiked rushes with apparently bleached tips wave in the wind like wheat and provide food and nesting material for magpie geese. The geese feed on the tuberous corms of the plant and use its foliage for nests; they also feed on the seeds of these rushes, and of wild rice.

The lowland areas of Kakadu support mixed scrub made up of many flowering species, as well as forest woodland and savannah grassland communities.

The forest woodland is dominated by eucalypts and gives way to more open woodland, then to the savannah grassland communities in progressively drier areas. Two eucalypts, woollybutts (*Eucalyptus miniata*) and stringybarks (*E. tetradonta*), are dominant in the woodlands, with tall shrub communities

Delicate eucalypt flowers and their hardy seedpods.

consisting of sand palms (*Livistona humilis* and *L. inermis*), *Grevillea* and *Acacia* forming a layer beneath the canopy. The same vegetation profile also occurs in the deep sands of the plateau region.

The dry woodland also contains the distinctive ghost gums, with their shimmering white or grey trunks, bloodwoods and ironwoods. Kakadu has twenty eight species of eucalypts, three of which have not been recorded elsewhere.

The cover is far more patchy, or non-existent in the savannah grasslands, but mixed scrub occurs on gentle lowland slopes. *Grevillea*, *Melaleuca* and some eucalypts can be found here.

The monsoon forest community appears in isolated patches on the edges of floodplains, along permanent freshwater watercourses and lagoons and in gorges at the base of the escarpment. The forest is basically non-eucalypt and needs permanent underground or surface freshwater for survival. An evergreen tree canopy, buttress roots and many species of lichen, vines, mosses, ferns and orchids thrive in the complex structure of the monsoon forest. Such areas are under constant threat from buffalo and pigs, which use the forest patches for Dry season refuge and camping.

One of the most fascinating categories of vegetation in the region is the rainforest of the escarpment and plateau sandstone, almost exclusively made up of *Allosyncarpia ternata*, a dark green tree with leathery leaves and a closed canopy. This evergreen myrtle tree, which grows to about thirty-five metres, was not scientifically described until 1976.

On the plateau large areas of heath and scrub cling to thin soil with better, deeper soil supporting stands of eucalypt woodland and a ground cover of spinifex and wiry grasses.

The huge diversity of Kakadu's flora gives forth a great variety of fruits and blooms, to be feasted on by birds, insects and mammals.

The Darwin woollybutt of the woodland areas, for example, has a striking red-orange blossom typical of the bright colouring of many Australian species. Another eucalypt, *Eucalyptus phoenicea*, displays its segmented orange and yellow blooms on the edge of the escarpment.

The cottonwood tree (*Bombax ceiba*) in the monsoon forests and along rivers has large yellow and red flowers with large, curled petals.

Shrubs and trees drape the escarpment, their flowers softening the harsh contours of its weathered rock.

Up to ten species of *Grevillea* have been recorded from the region. The flowers of these trees and prostrate shrubs have orange bottlebrush flowers, or red or yellow tentacles of blossom.

The creamy blossoms of ghost gums and other eucalypts attract many birds, insects and flying foxes and *Melaleuca, Acacia, Hibiscus* and *Cassia* also produce colourful scented flowers. *Gardenia*, common in the north of the park, has one of the sweetest flowers and one of the brightest displays is the purple and white pronged bloom of the turkey bush.

In the wetter areas of the park the water lily *Nymphoides hydrocharoides* weaves a yellow carpet around the edges of billabongs, and the lotus lilies *Nelumbo nucifera* and waterlily *Nymphaea gigantea* enhance the classic billabong scenes with red, white and blue flowers.

Several deciduous trees in the region display bright flowers in the Dry season. Kapoks (*Cochlospermum fraseri*) have a dazzling yellow, large petalled flower; kurrajongs (*Brachychiton paradoxum*) have a red trumpet shaped flower. The kurrajong tree is one of the many plants used by Aborigines: the fruit is eaten, the bark used for bags and other materials and the trunk for dugout canoes. Kapok was used for decoration in dances and rituals.

The plants of Kakadu provide Aborigines with many kinds of food, fruit, roots, seeds and nuts, leaves for spicing food and flavouring. Trees are also used for painting, firewood, weapons, musical instruments, trap making and for medicine.

Left: Hibiscus is one of the more common flowers of Kakadu.
Overleaf: Grevillea, one of Kakadu's most prolific plants.

Fungi appear on the forest floor after rain, to die away as the
moisture evaporates.

Spears of native grass float on a billabong.

Above: *Lindernia,* a colourful bloom on the floodplains.

Left: Grevillea blossoms attract native birds to their rich supplies
of nectar.

Overleaf: A tropical downpour has just ended, the water is rapidly
draining away, and the pandanus responds with lush green growth.

Above: Lilies in flower amid the reeds of the swamplands.

Left: A selection of Kakadu's prolific flora.

Overleaf: A delicate lily flowers in the shelter of a crevice in the
escarpment rock.

Moss and lichen combine with the elements to erode the
weathered wood of a venerable tree.

With the coming of rain, small aquatic plants flourish in a
temporary rock pool. As the water dwindles, the plants too
will dry up and die off.

Above: The blossoms of a native grevillea, one of several species found in Kakadu.

Left: Protected by the canopy of the monsoonal rainforest, plants respond to the rains with a profusion of blossoms.

Overleaf: A hardy grevillea flourishes in the rocky terrain at the base of the escarpment.

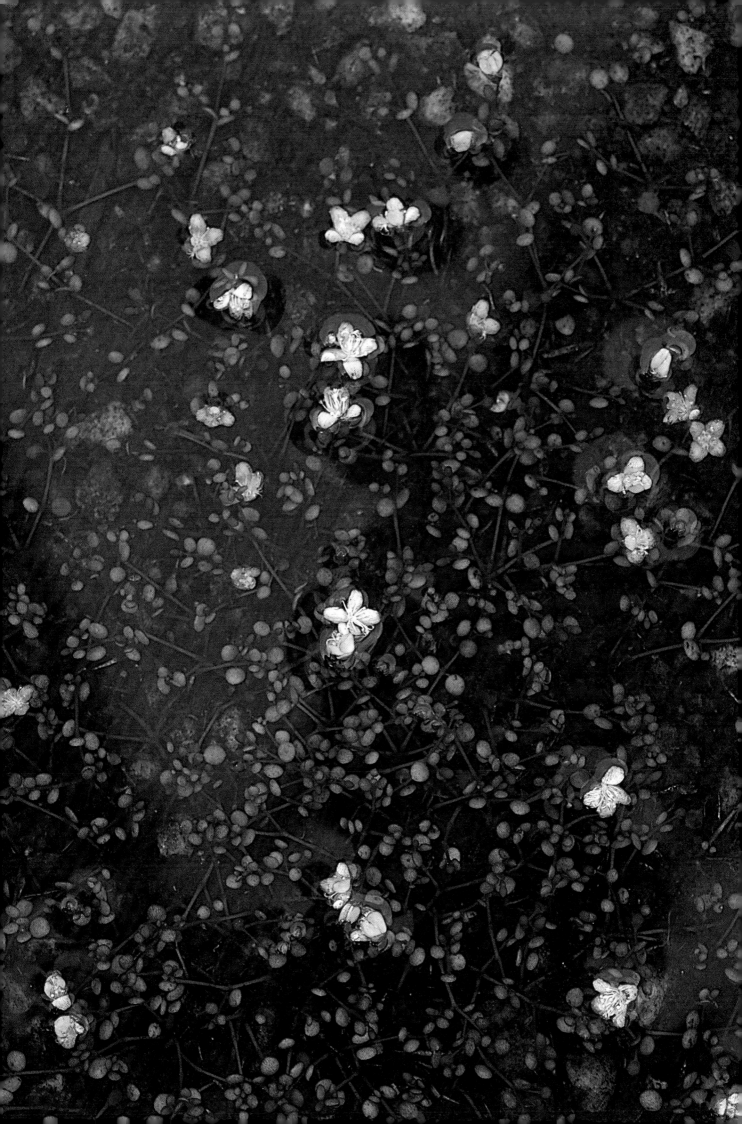

Above: *Cleome viscosa*, one of Kakadu's many colourful blooms.

Left: Blossoms of many kinds flourish in the aftermath of rain.

Top: *Hibbertia* blossoms after Wet season rains.

Above: A native hibiscus.

Right: *Passiflora foetida*, the climbing wild passionfruit.

Above: The damp monsoonal floor is the perfect habitat for a variety of fungi that appear amid the leaf litter and decaying foliage.

Left: The hardy pandanus seed is adapted to shoot even in the hostile environment of seemingly barren rock.

Overleaf: It is evening and the monsoonal rains are not far away, but the billabong, with its flowering lilies, is now calm and peaceful.

Above: *Sowerbaea alliacea,* freshly budding in the open woodland.

Left: *Carthonena* attracts native bees and insects.

Overleaf: Flowering grasses fringe a billabong after rain.

# WILDERNESS SANCTUARY

## A HOME TO SPECTACULAR WILDLIFE

Above: Kakadu's five species of geckos
hunt insects at night.

Left: A female darter or snake-bird
(*Anhinga melanogaster*).

## INSECTS

The insects of Kakadu are by far the most populous creatures of the region. In a 1973 survey, a collection of 65 000 specimens revealed 4800 species, little more than half the actual diversity in the park: there are an estimated 4000 species yet to be recorded.

Sustained high temperatures and a diversity of habitats support huge numbers and varieties. More than 1500 species of moths and butterflies and an estimated 1350 beetles were collected in 1973: there were even seventy-two types of dragonflies and at least another fifteen are known to occur in the region. Clouds of adult dragonflies herald the end of the Wet and at least two species of dragonflies with southern affinities have, surprisingly, been found in Kakadu.

Most species of insects are restricted to three distinctive and restricted habitats: large, permanent waterholes supporting beetles, midges, dragonflies and other species needing standing water during the Dry; moist forest areas; and permanent spring-fed streams.

Many species move from open woodland to moist patches of rainforest for refuge in the Dry, creating huge clouds.

There are more than 600 species of ants, wasps and bees, some of which are found nowhere else. Green ants swarm throughout the park, binding together leaves in trees as nests and attacking any disturbers of their peace.

A survey of mangrove forests and intertidal zones in 1979 recorded several invertebrate species new to science, including a new species of wolf spider, one of twenty-six arachnids discovered in the survey, which also found thirty-two species of crabs and twenty-six snails.

There are also fifty or so species of termites, including giant termites, found only in northern tropical Australia, that are the sole survivors of a family otherwise known by 200-million-year-old fossils from other continents.

Other species of termites build mounds in the region, and the grass-eating termites' mounds may be more than three metres high.

When Leichhardt's team passed through the region in 1845, it identified and named a magnificent blue and red grasshopper nearly six centimetres long. The Leichhardt's grasshopper, restricted to a few locations because of its specific feeding habits, was rediscovered only in 1973.

Leichhardt's grasshopper, first collected by Ludwig Leichhardt but rediscovered only in the 1970s.

The birth of a cicada. It has broken open its hard casing and
is pumping up its still crumpled wings.

A brightly coloured and patterned moth on a weathered tree.

## FISHES

There are more than fifty species of fish in the region, which boasts a remarkable diversity of freshwater species. By 1980 twenty-eight species had been recorded in the Magela and Nourlangie Creeks, twenty-two species in a small tributary of Deaf Adder Creek and fourteen species in the headwaters of Baroalba Creek. The Murray-Darling, the most extensive river system in Australia, has only twenty-seven native species.

The fishes of Kakadu, together with several species of reptiles and amphibians, have been important food sources for the Aborigines, as indicated by the many paintings of them in rock art.

Many species, such as the archer fish, the tailed sole, the one-gilled eel, the small-eyed sleeper, the Gilbert's grunter and the sharp-nosed grunter are extremely rare. The archer fish is thought to be restricted in Australia to the South Alligator River catchment, living in waterholes and springs near the escarpment and catching its insect prey by spitting drops of water for distances of up to a metre and a half.

Another uncommon species, the saratoga, is one of eight species in the world belonging to an otherwise extinct family. The one-gilled eel is capable of aerial respiration and the freshwater longtom is one of only two freshwater species.

The best-known of Kakadu's fish is the silver barramundi or giant perch, one of at least five species of sporting fish in the area. It has been commercially exploited, and has become part of the folklore and image of the Top End. Barramundi have been over-exploited by visiting fishermen and strict controls are now enforced to reduce exploitation to a sustainable level. One fascinating aspect of the life of the barramundi is that it undergoes a sexual inversion; juveniles mature initially as males but become females in their fifth or sixth year.

## AMPHIBIANS AND REPTILES

The attractive, colourful and profuse butterflies represent the romantic, the beautifully benign side of nature.

But there is another side. One of the most important animals of Kakadu is the saltwater crocodile, a far cry from the region's lovely insects and birdlife. Environmental philosopher Val Plumwood has a romantic view of her profession. But her experience of nature became real in February 1986, when she was

Armies of green ants devour a grub, contributing to the endless nutrient cycle of Kakadu.

mauled and pulled underwater by a three-metre crocodile in the East Alligator River. During the struggle, she looked into its 'beautiful, flecked golden eyes', and was later angry at the shot that ended the wild life of a crocodile – there was little evidence it was the 'culprit'.

In March 1987, Kakadu crocodiles killed one careless person who had ignored both oral and written warnings. The crocodile had to be shot, to force it to let go of its victim's body. Plumwood is adamantly opposed to the 'taming' of the land. 'We are not always in control and perhaps should not be,' she says. 'I would like to see a part of the world unmanaged so there will still be wild places and wild animals. I don't want to see places like Kakadu become a managed farm; that's not what I want to see of the natural world.'

Kakadu is one area of the world where the 'wild' is managed, if only to protect it. Should the management be kept to a minimum? Some would say it is, pointing to the principle of conservative, protective habitat management and the priorities of removing alien species such as buffalo and mimosa in order to preserve the natural order and natural diversity. But someone forgot to tell the crocodiles, which, in some areas of Australia, have been killed senselessly.

Whatever the answer, there is an undoubtedly a wilder, dark side to Kakadu, in which the raw power of such a region is manifest. And there is no doubt that such power and destruction are not only perfectly natural, but that they must be treated with the 'respect they deserve', as the Conservation Commission of the Northern Territory points out in a pamphlet on crocodiles.

All the main Australian groups of reptiles are represented in Kakadu; the benign climate presents few difficulties in maintaining a stable body temperature. The great contrast between the Wet and Dry seasons is of more concern to the amphibians and reptiles and seasonal conditions of inundation, high and low moisture, and salinity are major factors in determining their distribution within the region. Many reptiles better suited to more arid conditions have problems of excess water in the Wet season in areas such as the blacksoil floodplains, which are seasonally inundated. In contrast, frogs prefer such conditions and experience more problems during the Dry. The need for fresh water for breeding determines the habitats and cycles of frogs, all of which breed during the Wet and become markedly less active to survive the Dry.

Tree frogs emerge from their shelters after rain; during the
Dry season they are rarely seen.

Above: A Gould's goanna, superbly camouflaged for its life among the rocks and trees.

Overleaf: A black-headed python suns itself on the rocks at the base of the escarpment.

One hundred species of amphibians and reptiles have been recorded from a broad range of habitats in Kakadu. Five species of geckos, twelve skinks, five elapids (venomous front-fanged snakes) and five worm snakes are widespread, together with three species of pythons, two dragon lizards, two colubrid (venomous rear-fanged) snakes, a monitor and one species of the highly evolved, snake-like flap-footed lizards.

Several species are found only or primarily in Kakadu. One of the larger snakes, the beautifully coloured Oenpelli python, *Python oenpelliensis* – restricted to the stony escarpment country – was discovered in 1977, and the large gecko, *Pseudothecadactylus lindneri*, is also restricted to these areas. Arnhem land skinks, first identified in 1981, are found only on the sandstone massifs and gorges of the escarpment. Two species of frogs are generally restricted to Kakadu as well.

Of the five or possibly more species of freshwater turtles found in the region, the most unusual is the Fly River, or pig-nosed turtle (also known as the pitted-shelled turtle, and the only member of the family Carettochelyidae), which hides in this vegetation in estuarine and freshwater environments. Although Kakadu's Aboriginal rock art features detailed drawings of this turtle, it was thought to be only confined to the south coast of Papua New Guinea until found in the Daly River in 1969. It was discovered in Kakadu in 1973 and is now believed to be widespread, though rarely seen, in Top End rivers. It is thought to eat pandanus fruit, leaves and molluscs, and may have multiple nestings during the Dry season.

The most common turtles are the northern snake-necked turtles, which are exclusively carnivorous and the snapping turtles, which are exclusively herbivorous (most freshwater turtles are omnivores).

Several species of marine turtles also nest on the beaches of Kakadu. The flat-back, green, hawksbill and Pacific ridley turtles live in the coastal waters after emerging from their nests on the beach and running the gauntlet of goannas and other predators. Many turtle eggs are also taken from the nests by predators.

The snakes of Kakadu are numerous and varied and common on the floodplains, in open eucalypt and monsoon forests and in the escarpment country, though they are rarely seen by humans.

Many rear-fanged snakes live among the mangroves, hunting crustaceans and fish; one species lives in freshwater, where it consumes frogs. This group, and the file snakes, inhabit the permanently inundated areas of Kakadu that are generally avoided by elapid (front-fanged) and boid (python) snakes. The water python, however, is common in freshwater billabongs and creeks.

In the unshaded sedgelands where seasonal inundation occurs, several species of snakes prey on frogs during the Wet. The water and olive pythons are common and are joined late in the Dry season, when nights are warm and humid, by death adders, one of the area's dangerous venomous snakes. Much of the danger from death adders stems from their habit of remaining perfectly still and striking (rather than retreating) when stepped on or passed.

King brown snakes and taipans are the other highly venomous snakes. The long (up to 2.5 metres), active and aggressive taipan is usually found in the escarpment areas, but the 2-metre-long king brown spends more time in wet areas, feeding on frogs and sometimes on the eggs of the magpie goose.

The most spectacularly colourful snakes are the pythons, central to several Aboriginal legends in Kakadu, the land of the Rainbow Serpent. Water pythons also eat the eggs of the magpie goose and hunt small birds and animals on the floodplains. Olive pythons, black headed pythons and Children's pythons stay more commonly in the woodlands, and the recently described Oenpelli python is endemic to the escarpment.

There are also many lizards in the Alligator Rivers region, some of which are at the northern limits of their wide distributions or in northern phases, like the blue-tongued lizard. The most striking are the dragon lizards, epitomised by the frilled lizard of the wooded regions. There are several more species of dragon lizards, but the frilled lizards, which extend a large membrane around their necks when surprised or threatened and run readily on their hind legs, are the most distinctive. But they are not the largest of the lizards: several species of goannas and monitors, including two freshwater monitors, are found in the region; the most common is probably the Gould's goanna.

Some of the smallest reptiles are the most commonly seen. Many species of insect-eating skinks scuttle over the ground litter and trees of the woodlands. At night the hunt is taken over by geckos, with their distinctive bulbous eyes and

With its mouth open and its neck extended, umbrella-like, the frill-necked lizard prepares to defend itself.

the padded feet they employ to scamper upside down across logs, rocks or ceilings. Two new species of skink and a new species of gecko have recently been discovered in the eucalypt forests of Kakadu.

The most spectacular of the reptiles are of course the crocodiles. *Crocodylus pororus*, the 'saltie' or saltwater crocodile, is as its name implies most common in estuarine environments; coastal swamps, tidal streams and mangroves. But this carnivorous species, which attacks large animals and has a well-developed territorial sense, also inhabits freshwater – even billabongs and creeks as far upstream as the escarpment base.

The saltwater crocodile can be distinguished by its broad snout, and is generally larger than the narrow-snouted freshwater crocodile *Crocodylus johnstoni*. Salties grow to five metres, whereas the freshwater crocodiles are usually less than three metres in length and are unlikely to attack large animals such as humans. They use their slender snouts to catch fishes, frogs and reptiles in the upland watercourses and billabongs.

The peak of the saltwater crocodiles' nesting season is the onset of the monsoon, when they build mounds of soil and vegetation near permanent water. Freshwater crocodiles, in contrast, nest in the Dry season, depositing their eggs in chambers on sandbars.

The crocodiles are at the apex of one of the region's food chains and a sustained population is vital to the balance of many species. Pollution of freshwater streams or swamps is a danger to aquatic reptiles either directly or through contamination of their food chain, and such contamination could affect crocodiles. Any leak or release of contamination from the Ranger uranium mine into Magela Creek could be catastrophic; Magela Creek supports a high density of large saltwater crocodiles, a situation not repeated anywhere else in northern Australia.

But the main reasons for the dramatic decline in the crocodile population is the historical harvesting by humans for commercial, or 'safety' reasons.

Hunting for skins, uncontrolled shooting and fish nets accounted for many thousands of crocodiles, though the species is now officially protected. Opinions vary as to the population's recovery, but one researcher estimates the present population of saltwater crocodiles to be thirty to fifty per cent of the original population of fewer than 100 000.

Above: A saltwater or estuarine crocodile slips into a billabong, watched by a flock of whistling ducks. The crocodile is now protected from indiscriminate hunting.

Overleaf: A clay termite mound stands in the open forests as monsoonal clouds billow on the horizon. Kakadu boasts fifty species of termites.

Although buffalo still have an adverse effect on the freshwater crocodile population by increasing salinity, culling of this destructive quadruped is succeeding and the slashing of barramundi fishing licences has reduced the mortality of non-target species such as crocodiles.

Some crocodiles have been relocated to avoid human fatalities, but their homing instincts are strong enough to enable them to return over long distances; capture can also lead to the death of these surprisingly delicate animals.

Close to the other end of the reptile and amphibian scale are the frogs of Kakadu, highly adapted to the region's severe climatic changes. Many are restricted to quite small habitats and lie dormant during the Dry. But at the onset of the Wet they appear in huge numbers to revel in the water and the burgeoning food supply, which includes insects, dragonfly nymphs, algae, weeds and tadpoles and fully grown frogs of other species.

Breeding takes place only during the Wet season, and the chorus of thousands of frogs around billabongs is designed to attract females. Many of the twenty-five species are, in fact, most easily identified and located through their calls. The green tree frog, for example, issues a booming call from tree trunks.

There are superbly coloured (often for camouflage) frogs of various sizes, from the large tree frogs to the dwarf rocket frogs. Their habits, too, differ markedly: *Litoria dahlii*, a poorly known species whose call has only recently been recorded, floats in the water with legs outstretched, only its eyes protruding above the surface as it preys on insects and at least four types of frog, including members of its own species.

### BIRDS

The billowing clouds are long gone and the air has lost its steam. With little likelihood of rain for months, the circle of Kadadu's year is closing. The park's spaces may be vast, but they are finite; late in the Dry season literally millions of birds are feeling that finiteness.

Where water and food remain the birds congregate, trapped by the encroaching Dry. A slow boat trip down the backwaters of the South Alligator reveals a dazzling variety of species as the water supply shrinks.

Among the irridescent foliage and lilied stretches of water, both shining under the June sun, species after species share the wetlands of Kakadu. From the night

Twenty-five species of frogs inhabit the swamps of Kakadu, each distinguished by colour and sound.

heron to the white-breasted sea eagle, the jacana and the yellow spoonbill, the forest, azure and sacred kingfishers and the bar-shouldered dove past thousands upon thousands of magpie geese, the display has an air of unreality, or organised intensity, about it. But there is no artificial entrapment keeping the birds within reach – only the vicissitudes of the seasons and the strict order of nature.

The concentration of species may be seen in the wetlands well into the Dry season, but in all its habitats and throughout the vagaries of its climate, Kakadu's birdlife is one of its most compelling features.

The birds appear singly or in pairs, like the rare white-throated grass wrens of the escarpment, or in millions, like the magpie geese, late in the Dry season. Collectively they form one of the most varied and spectacular bird populations found anywhere in the world: in all, 275 species of bird have been recorded at Kakadu; three are found nowhere else.

The draining of wetlands throughout the world is keeping pace with the equally destructive denudation of forests as tropical freshwater swamps and plains are lost at the hands of agriculture or irreparable damage upstream.

The world's forests, with their intense diversity and oxygen-producing prowess, have exercised the minds of those concerned with the world's survival. But the planet's wetlands are just as important, for these vast areas play an equally vital role in keeping intact the evolutionary chain, the diversity of species and the future of the natural world. Major disturbance to any habitat affects the wildlife of that region . . . and as the wetlands are such an integral part of Kakadu, the entire river system must remain intact for the birds to survive.

When Ludwig Leichhardt passed through the region in 1845, he noted the abundant birdlife in his diaries, recognising it as an important part of the natural structure. He wrote of 'innumerable geese, ducks, native companions, white cranes and various other waterfowl'.

Now, Kakadu's wetlands are on the United Nations list of Wetlands of International Importance especially as Waterfowl Habitat; 2700 square kilometres are either seasonally or permanently inundated, and many millions of birds depend on these coastal swamps, tidal flats and floodplains.

The mangroves of the coastal and riverine fringes, close to or at the end of the rivers' journeys from the escarpment, provide sanctuary and important roosting

A darter or snake-bird, one of Kakadu's many waterbirds,
dries its wings on a rotting tree trunk.

and breeding sites for many species. Some, such as egrets and cormorants, use the mangroves for their nesting colonies and spread out over the wetlands.

The large-billed gerygone (also known as the large-billed warbler), the mangrove gerygone or mangrove warbler and the chestnut rail are some of the species found only in the mangroves. These biologically rich areas, where twenty-one of the twenty-nine Australian mangrove species occur, are also the exclusive habitats of birds such as the mangrove kingfisher, the broad-billed flycatcher, the red-headed honeyeater, the mangrove robin and the white-breasted whistler.

The tiny yellow silvereye or white-eye can be seen in the mangroves or rainforests, searching for insects, berries and seeds. The nest of the silver-eye is an exquisite work of art, a small cup of fibrous materials suspended in a mangrove. The mangrove or buff-breasted warbler builds superb dome-shaped nests with hooded entrances, also made of fibrous materials such as seaweed and bark and suspended from branches.

These nests contrast with those of egrets and cormorants, which use the mangroves for their large nesting colonies and which build bulky structures and platforms from sticks.

The wetlands of Kakadu also see many migratory species, and the coastal and mangrove fringe has a spectacularly varied population. Herons, egrets, night herons, jabiru storks, ibises, ducks, ospreys, kites, white-breasted sea eagles, plovers, dotterels, turnstones, curlews, sandpipers, terns and gulls, kingfishers, martins and cuckoos, thrushes, flycatchers and whistlers all frequent the fringe.

Other common birds include the white-browed crake, the darter, the brown booby or gannet, the greater and lesser frigatebirds, the brahminy kite, the grey-tailed tattler, the Torres Strait pigeon and the bar-shouldered dove, the rainbow bee-eater and the mistletoe bird. Ospreys glide overhead, ready to dive and to seize fish for their nestlings.

The wetlands of Kakadu, especially the floodplains and their expanses of sedge and grass, constitute one of Australia's few refuges of this type and are certainly Australia's major refuge for many tropical waterbirds.

*Anseranas semipalmata*, the magpie goose, is one such species. Because of the enormous wetland flocks that congregate each year, this bird is sometimes considered abundant beyond danger. But the magpie goose's habitat is shrinking:

Above: Flocks of magpie geese constantly descend on the billabongs in search of food and water.

Overleaf: The lotus bird or lilytrotter has long back toes to enable it to walk delicately on water foliage. It nests on a platform of foliage on a billabong.

Kakadu's grasses and sedges provide a Dry season refuge for the geese, which are now generally confined to northern Australia. The species was once widespread in Australia, and as one of the most important wildfowl in Australia its diminution is a cause for international and domestic concern.

The magpie goose is the sole member of the subfamily Anseranatinae, and is by no means safe. Research has suggested that if the noxious weed mimosa is not controlled within ten years magpie geese would probably be classified as endangered, so management control of mimosa is vital. While the addition of Stage Two to Kakadu National Park provided an extensive floodplain refuge for these birds, the threat from mimosa and buffalo is still present.

In the Dry season most geese are concentrated in the Boggy Plains-Nourlangie region, the Northern Territory's major Dry season refuge, where they depend on freshwater and Dry season food from stands of the spiked rush, *Eleocharis dulcis*. The foliage of this spiky plant is used by the geese for nests and they feed on its tuberous corms and seeds.

If the Wet season is late and these plains dry up, the Magela Creek plains and upper reaches of Nourlangie Creek gain even greater importance: research statistics show that in the 1983 Dry season, more than two million geese from a Northern Territory population of fewer than three million were observed on the National Park's floodplains. In 1984, 2.5 million of 3.9 million were in the park.

Thirty-five species of waders have been recorded in the wetlands, many of them migrants using Kakadu as their landing place in Australia. The floodplains also support large numbers of pied heron, plumed and wandering whistling ducks, Burdekin or radjah shelducks, glossy and straw-necked ibises, jabiru storks and the green pygmy goose. These species are largely confined to northern Australia but are joined by more widespread species such as Pacific black ducks, egrets, night herons, white-faced herons, pelicans and darters.

The jabiru stork is one of the most exotic, yet attractive, birds of the park. Its huge red legs seem to shorten as it wades into the water, looking for fish and frogs to spear with its long, sword-like beak.

Another fascinating species is the comb-crested jacana, or lotusbird. This small bird relies on its elongated toes to spread its weight across the leaves of lilies and other vegetation as it dances across swamps looking for food.

Above left: A jabiru scoops up water from the marshlands.
Below left: The white-breasted sea eagle is a superb predator.

The brolga, or 'native companion', a member of the crane family, is also plentiful on the floodplains, inland waters, fringing forests and open woodlands of the region. This large grey bird moves gracefully and engages in superb dancing routines with prospective mates.

In the open, treeless sedgelands and grasslands, other non-aquatic birds make their homes. The tawny grassbird, zitting cisticola and the golden-headed cisticola – all members of the Old World warbler family – are common on the plains and the crimson finch, the chestnut-breasted and yellow-rumped mannikins and the rare yellow chat are also found there.

A boat trip along the waterways of the floodplains may reveal the artfully camouflaged shapes of Nankeen or rufous night herons lurking in the undergrowth. They remain perfectly still for long periods as they wait patiently for prey to emerge. In 1981, the only known Northern Territory rookery of these birds was discovered deep in an almost inaccessible rainforest near the mouth of the South Alligator River. Two thousand of these shy herons nest in a rookery no more than 400 metres long by 200 metres wide.

Kakadu also contains several rare species unique to or largely restricted to the Kakadu region. Most can be found only on the Arnhem Land plateau and the escarpment or its outliers, sheltering and feeding in the forests and gorges.

The banded fruit-dove or black-banded pigeon (*Ptilinopus cinctus*) is sparsely populated throughout the escarpment, and has also been sighted near the mouth of the South Alligator River. It feeds on the canopy fruits of the patches of relict rainforest in the gorges.

The chestnut-quilled rock pigeon (*Petrophassa rufipennis*) stays on the sandstone slabs and screes of the escarpment, feeding on seeds, but can sometimes be seen at outliers such as Nourlangie Rock.

Another unique Kakadu species, the white-throated grasswren, is found in the spinifex vegetation of the exposed, flat top of the plateau. A rare lavender-flanked form of the variegated wren, *Malurus lamberti*, also frequents this area.

Many species of birds range over the variety of habitats in Kakadu, moving according to seasonal factors such as rainfall, the blossoming and seeding of trees and the frequency of fire. The white-lined honeyeater, in contrast, is confined to the gorges of the escarpment and until recently was considered unique

The tiny double-barred finch is an agile seed-eater, feeding
on the grasses of the floodplains.

to the park (another small population has recently been found in the Kimberley area of Western Australia). This bird is restricted in its habitat to the sandstone rainforest or the broadleaf shrubbery associated with it, as are the banded fruit-dove and the sandstone or helmeted friar-bird (*Philemon buceroides*).

The rare red goshawk, the sandstone shrike-thrush and the hooded parrot (*Psephotis dissimilis*) inhabit the open forests and hill woodlands of the plateau and escarpment. The little-known hooded parrot is an endangered species that ventures into eucalypt forests to feed.

The rainforests of Kakadu also support an important bird population. Isolated pockets of sandstone and coastal deciduous rainforests have permanent residents, such as the rufous owl and the nankeen night-heron that use them solely for roosting, and birds that visit them for food.

Water buffalo have had a substantial effect on some species in the rainforests. The orange-footed scrubfowl, a ground-nesting and ground-insect feeding bird, the rainbow pitta and the spangled drongo are among the casualties of buffalo damage to fragile rainforest habitats.

The grey whistler is found only in the mangroves and the coastal rainforest. The little shrike-thrush, the cicadabird (a cuckoo-shrike) and the green-backed warbler or gerygone, one of the five Australian warblers found in the park, frequent the coastal and sandstone pockets of rainforest.

The open woodlands and other forests also have a large bird population that encompasses a variety of species. A number of species reside in the paperbark forests, and more feed when the *Melaleuca* trees are flowering. At such times honeyeaters, red-collared and varied lorikeets and several parrots, along with other nectar and insect-eating birds, feed on the blossoms and the animals they attract. The brush cuckoo, white-browed robin and lemon-breasted, restless and paperbark flycatchers remain in these forests throughout the year.

In the open woodlands the brilliantly coloured Gouldian finch struggles for survival. It is rare and populations are declining, and its survival in the wild is dependent on the refuges in the South Alligator River catchment.

The crested hawk and the square-tailed kite are two birds of prey that watch over the lowland woodland and open forest and, by their very nature, range over most of the region together with eagles, falcons and the osprey.

A blue-winged kookaburra, with its catch secure. This species is well suited to the Kakadu habitat.

Perhaps the most regal and awesome of the birds of prey are the wedge-tailed eagle and the white-breasted sea-eagle, which perch high in dead trees along the waterways. The grey or white goshawks have a lower, more wooded habitat than their red relatives and the smaller kestrels, sparrowhawks and falcons frequent the various wooded habitats of Kakadu. What the sight of the black kite is to the park, the sound of the whistling kite is to this region: both are prolific, unlike the rare letter-winged kite that is found only on the floodplains.

The red-tailed black cockatoo, the sulphur-crested cockatoo, the galah and the busy lorikeets make their presence known by their raucous calls. The related little corella frequents the higher, drier areas, like the tawny frogmouth, the owlet nightjar, the emu and bustard. Owls and cuckoos are also found in many areas of the park, though the bustard, the great bowerbird and several species of butcherbird of the floodplains and open woodlands are increasingly rare.

Buffalo and fire are constant pressures on the region's bird populations; grass finches and the hooded parrot are particularly vulnerable to fire in their specialised habitats. Waterfowl hunters, unregistered in the Northern Territory, and smugglers of parrots and finches are also a danger to the birdlife that is a vital and attractive part of Kakadu.

## MAMMALS

The mammals of Kakadu form a diverse group of at least seventeen marsupials, twenty-six bats, fourteen rodents and one monotreme, the echidna.

Most are found in the open forests and woodlands and on the escarpment and plateau, while some occur on the floodplains and in the rainforests.

Twenty-six of Australia's sixty-five species of bats are found in Kakadu. Thousands of fruit-bats or flying foxes disperse from their forest colonies at dusk to feed on fruits and blossoms. Smaller bats, such as the northern blossom bats, live in escarpment caves or hollow trees.

Kakadu has several rare or endangered species of bats, including the lesser wartnosed horseshoe bat, the diadem horseshoe bat and the orange horseshoe bat and the ghost bat, Australia's largest insect-eating bat. White-striped sheathtail bats were discovered in the park in 1979.

The kangaroos and wallabies are also well represented, with agile wallabies feeding in groups on the open plains and woodlands. Wallaroos prefer their own

Agile wallabies are among the most common macropods of Kakadu, at home in the open woodland and grasslands.

company in wooded areas. A rarer macropod, the antilopine kangaroo (a northern relative of the red kangaroo) frequents ridges and woodlands. The black wallaroo, the little rock-wallaby and Harney's *Antechinus* (discovered on the escarpment in 1948) and fawn marsupial-mice are dependent on the park for their survival.

The northern quoll or native cat is a distinctive inhabitant of the area. This marsupial carnivore hunts fiercely and busily at night on the escarpment.

Some native rodents, among them the false water-rat, the golden-backed tree-rat and Woodward's rock-rat, are also restricted to or are most populous within the region.

The park is also home to possums, sugar gliders and bandicoots in wooded areas, and some species of rats and possums join the wallabies, nabarleks, echidnas and marsupial mice on the rocky escarpments and outliers.

Completing the mammal fauna of Kakadu is the dingo, *Canis familiaris*, which is widespread in the park and which feeds on small animals such as birds and reptiles.

The dingo has remained generally purebred in the region and superb gold specimens can be seen hunting singly, in pairs or in small groups.

Northern Territory tourist advertisements show buffalo charging through floodplains and proclaim that the Territory has 'a wildlife all of its own'. But there are thousands upon thousands of native species in Kakadu that have suffered from the damage caused by the introduced buffalo. These native animals are the wildlife Kakadu can truly call 'its own'.

Left: The phascogale, a fierce and agile nocturnal marsupial, feeds on insects and other small creatures.

Overleaf: The dingoes of Kakadu move with the cautious confidence of creatures in command of their habitat.

# TIMELESS CULTURE

## ABORIGINAL SETTLEMENT AND ART

Above: The dwindling waters of a
billabong near the East Alligator River.

Left: Aboriginal cave painting, testimony
to an ancient civilisation.

People throughout the world worship their perceived creators, believing that all life emanated from them. Creators take many forms, amorphous or absolute, and can be as great and real as mountains. The world's highest mountain is *Qomolungma* – 'Mother Goddess of the Earth' – to the Tibetans who live at her feet.

In the Kakadu region the Great Earth Mother is also of integral importance, and a major theme in Aboriginal beliefs about Ancestor beings. Naturally, she is a symbol of fertility and the creator of life. Sacred ceremonies in her honour ensure the reproduction of species and give life and energy to humans and animals.

Moreover, her importance to the Gagudju goes beyond fertility to a responsibility for the fluctuation of the seasons and the coming and going of rains. From her came all things and in her hands lie the fortunes of the Aboriginal world.

To the Gagudju the Great Earth Mother was Imberombera. She came across the sea, arriving on the coast of Arnhem Land, and was the bearer of riches and children. Children were in her womb and dilly bags suspended from her head contained yams, bulbs and tubers. In one hand she held a digging stick and as she journeyed across Western Arnhem Land she planted yams, palms, bamboo and waterlilies. She also built hills and creeks, flora and fauna and left on the land her spirit children, groups of whom were given different langauges.

Ungulla, another fertility Mother, travelled the northern country some time after Imberombera, meeting the children as she went. She wore paperbark sheets and showed the Gagudju women how to make bark aprons. After bearing many children she is said to have torn out her vagina and uterus, throwing them to the women and saying 'From now on this will be yours. You can have the children from now on'. She also gave her breasts and fighting-stick to the women, and a flat spear-thrower and a reed spear to the men.

The Gunwinggu people of Arnhem Land also tell a story of their Mother arriving from across the seas. Waramurungundji, as they call her, came from the northwest, landing on the coast at the very beginning of Creation. According to Gunwinggu tradition Waramurungundji – the "mother" – came from over the sea from the northwest, in the direction of Indonesia, at the beginning of the world. When she landed on the Australian coast she made children, telling them where they were to live and what language they were to speak. She also created

The rocks and trees, the land itself, are of immense
spiritual significance to Kakadu's Aboriginal people.

much of the countryside and left various creatures and natural features; bees and wild honey in one place and a banyan tree in another. She tried to circumcise the children she had made but at first she was unsuccessful and the children died. In those areas, people do not practise circumcision today. But at last she succeeded and in those places people continue to circumcise today. Waramurungundji's husband was called Wuragag, and he too came with her. After some adventures together he left her, and took a second wife named Goringal. Waramurungundji continued her journeys alone. Wuragag had many adventures, and many wives. At the end of his particular travels he turned into a high rocky hill, a landmark that dominates the plains north of Oenpelli. This bears his name, Wuragag, because his spirit remains there; in English it is called Tor Rock. Beside this rock stands a smaller one facing east and that is his young wife Goringal.'

These legendary ancestors not only came from across the sea, but by canoe. Science has arrived at similar conclusions from archaeological and anthropological evidence as prehistorians sift through the remnants of more than 40 000 years of human habitation. When Moses led his followers from Egypt, Aboriginal occupation of the Alligator Rivers region had been continuous for more than 20 000 years, and Aboriginal myths are rich in symbolic material that spells out the history of the Kakadu region.

Material manifestations of the oral traditions and symbolism enshrined in the Old Testament have been revealed by archaeology, and similar work in Australia has uncovered striking evidence of historic events described in the tales of Imberombera, Waramurungundji and other creator ancestors.

The first Australians may have been attracted to the new land by smoke from natural fires in the vast savannah to the southeast, or were driven on by the northwest monsoon. At a time when sea levels were low and land plentiful, it is hardly likely that population pressures sent them in search of new territory. It may not even have been an epic voyage, merely an extension of leaps between island stepping stones or the culmination of a chance voyage by a small group blown off-course during a routine journey from one island to another.

It is likely the journey was made when the sea barrier between the southern continent and Asia was at or near its narrowest, so world temperatures, glacial extents and sea levels are clues to the arrival of the first Australians.

IAN MORRIS

Above: Lightning strikes, heralding another tropical storm over Kakadu.

Overleaf: Rocky outcrops are a feature of much of the Kakadu landscape,
where the escarpment meets the floodplains below.

In the past 50 000 to 55 000 years there have been two periods during which sea levels have been more than 100 metres lower than they are today. Both occurred during the Pleistocene epoch, which ended about 10 000 years ago.

The two major reductions in sea level – to between 120 and 150 metres below present levels – occurred about 53 000 and 20 000 or 18 000 years ago, when the most recent glacial maximum occurred. But 18 000 years ago Aboriginal habitation of Australia was already widespread. Evidence at Lake Mungo in western New South Wales shows occupation for at least 10 000 years prior to the most recent glacial maximum. The cremated skeleton of a woman found at Lake Mungo has been carbon-dated at about 26 000 years old and further excavations have discovered stone tools that extend human activity in the area by at least another 10 000 years.

So it is likely that the first Australians came as the Dreamtime legends depict by canoe from Southeast Asia to a greatly extended southern continent, perhaps 50 000 years ago.

Australia would have been a third larger than it is today, probably wetter, more heavily vegetated and more fertile over a wider area, and abounding with wildlife. The first arrivals may have landed on a coastline up to 500 kilometres northwest of the present shore. Kakadu was at that time probably part of a vast inland plain, drier than it is today and covered in low woodland. The earliest campsite so far discovered in Northern Australia is at Malangangerr Nawamoyn, near Oenpelli in the Alligator Rivers region, with stone tools dating back beyond 20 000 years. The most significant of these tools were edge-ground (rather than flaked or knapped) axes, probably the oldest recorded evidence in the world of such an advanced stone tool.

The nearest comparisons with these ground axes in near-Pleistocene contexts are found in Papua New Guinea; the early colonists may have brought the techniques with them from the north.

Grindstones for flour making discovered in the Kakadu region greatly predate those of the Middle East, long thought to be the home of sedentary agriculture.

The invention or adoption of the grindstone by the Aborigines of the Alligator River between fifteen and twenty thousand years ago may well have enabled the intensification of inland occupation: 23 000-year-old grinding tools have been discovered in caverns by the East Alligator River at Oenpelli and a grindstone has also been found in the Alligator River area with some pigment used for decoration ground into it some 19 000 years ago.

Debate is intense as to when the inland of Australia was occupied. John Mulvaney, author of *The Prehistory of Australia*, assumes Australia was populated from the north or the northwest, with arrivals fanning out across the continent in a fashion similar to that spelled out in the creation stories told by Alligator River Aborigines. Such a theory would make the Gagudju Aborigines and their regional neighbours the first Australians – and the cultural and natural history of the Kakadu region indeed a window into the history of Australia.

The Oenpelli campsite contains food debris demonstrating that the diet of the early arrivals included turtles, bandicoots, possums, fish, shellfish, nuts and the roots of the lotus lily – foods still abundant in the area. The sophisticated axes, with their well shaped and sharpened edges, would have been used to chop wood and shape wooden tools while the stone flakes, also abundant at the site, were probably used for cutting vegetables, meat and for scraping hides.

The landscape and environment of the Kakadu region have witnessed many changes. Climate, vegetation and surface area have altered dramatically and the consequences of long-term climatic changes on the vegetation and wildlife have been considerable. Some wildlife species have become extinct; others, such as the Fly River turtle, have been thought to be eliminated. A painting in the X-ray style of this unusual turtle has been found at Little Nourlangie Rock, but the animal was only recently discovered by science in Australia. A specimen was collected in the Daly River region of the Northern Territory in 1969, and another was discovered in the Kakadu region during a government study.

The Tasmanian Tiger, or thylacine (*Thylacinus cynocephalus*), is known to have existed in the area. There are five identifiable paintings of this dog-sized predator with distinctive tapering hindquarters and tiger-like stripes in Arnhem Land. The Tasmanian Tiger is now thought to be extinct, though there are unconfirmed sightings in Tasmania to this day. The Tasmanian Devil (*Sarcophilus harrisii*) is known to have inhabited the Kakadu region: a 3000-year-old specimen has been discovered in Padypadiy rock shelter near Oenpelli, more than 2000 kilometres from its nearest present-day occurrence.

Above: Pools of water still remain from an earlier storm; another is
pending as clouds build up over the escarpment.

Overleaf: The marshy sedgelands and lush new growth are evidence of
the Wet season; in the Dry this land will be brown and vegetation will
be sparse.

Some species became extinct during Aboriginal occupancy. Habitat and environment altered due to major climatic shifts, with saline mud flats being turned into freshwater swamps and billabongs less than 1500 years ago, and hunting also altered animal populations and plant distribution.

'Statements that Aborigines did nothing to improve on nature are totally wrong,' says John Mulvaney. 'Fire-stick farming' (a phrase coined by Rhys Jones of the Australian National University) was carried out in the area, and this system of agriculture has demonstrated that Aborigines set out deliberately to develop and maintain stable food supplies. They fired the countryside to promote new grass for grazing. Fire made travelling easier, encouraged larger numbers of animals and increased the number and variety of food plants.

Archaeological discoveries show that the lives of Kakadu's Aboriginal population were dominated by the seasons. There was undoubtedly an abundance of game and food on the fertile plains during the Dry season, and hunters roamed freely. Flooding during the Wet season forced the population to move to the sandstone outliers or to the edge of the escarpment country to the east, where the use of rock shelters was widespread.

By moving with the seasons the Aboriginal population ensured that supplies of food would be regulated carefully and naturally. They discerned six main seasons in the year, using a calendar so finely tuned that it is still employed in the management of Kakadu National Park, determining the timing of such operations as controlled burning.

Kakadu's is essentially a tropical monsoon climate. From May to September the Dry season prevails, the plains teem with wildlife and the water systems provide rich food supplies for hundreds of species of birds. By October clouds are gathering and the humidity and temperature, high throughout the year, increase. Lightning and thunderclouds dominate the horizon. The region has an average annual rainfall of 1400 millimetres, virtually all of which falls between November and March. Spice is added by passing cyclones. At least nine crossed the area in the sixty years to 1975, and a further eleven passed nearby.

The monsoon climate dictates human movement as it has done for thousands of years. Tourist activity is normal  and most comfortable from May to August when the dry, cooler conditions prevail and travel is easier.

## THE ABORIGINAL ART OF KAKADU

At Ubirr near the East Alligator River, in one of the hundreds of rock shelters used by the Gagudju Aborigines for thousands of years, a convex boulder surface forms the ceiling of a magnificent art gallery.

The gallery can only be approached by wriggling and squeezing into a narrow crevice, from which a collection of Aboriginal art as deep as the culture it represents can be appreciated, not merely as a row of paintings in the pattern of a modern art gallery, but as a three-dimensional gallery. Paintings over paintings, beneath them more layers of artwork. The backbone of a barramundi is superimposed on the simple figures of other animals painted perhaps 15 000 years ago. Beneath that hand prints, or perhaps grass pressed against the rock prior to the last Ice Age, could well be covered by the work of later masters intent on telling legends, beckoning spirits or simply describing the day's experiences.

The paintings are no more than the outward expression of a massif of religion, legend, a caring, practical relationship with the land and a continuing interaction that has added to the culture rather than eroding it. Today, that cultural heritage is a prime quality of the Kakadu region. Rich beyond European Australians' belief, it reflects the Aboriginal view that the land owns the people, not the people the land. Kakadu experienced enormous changes during Aboriginal occupation; changes intertwined with Dreamtime stories and legends, manifest in their lifestyles and stored in the archaeological and art collections.

Sacred sites occur throughout the park and surrounding regions, including the area to the south, the Conservation Zone, that was to have been included in Stage Three of Kakadu National Park before the Federal Government granted permission for exploration in 30 per cent of the Gimbat and Goodparla pastoral leases.

In November 1845, Ludwig Leichhardt and his party had no inkling that they had stumbled upon what is probably the world's oldest collection of early artistic creativity. As Leichhardt's party entered the southern parts of Arnhem Land on its way north to the Cobourg Peninsula, Calvert, a member of the party, found a shelter. 'The remains of freshwater turtles were frequently noticed in the camps of the natives; and Mr Calvert had seen one depicted on the rocks. It is probable that this animal forms a considerable part of the food of the natives,' was the laconic entry in Leichhardt's journal.

A decorated reverse hand stencil, still clearly delineated in a rock shelter.

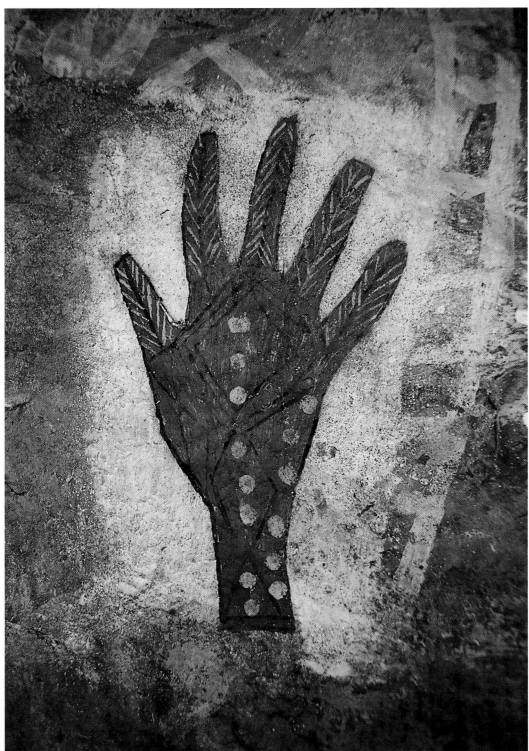

A recent X-ray style 'blue painting' at Burrunguy (Nourlangie Rock), by
Najombolmi (Barramundi Charlie).

In 1891, forty shelters containing about 200 paintings were discovered in the Alligator Rivers area and Dashwood, the Government Resident, reported paintings near Oenpelli in 1893. Baldwin Spencer visited Oenpelli in 1912 and described the rock paintings in the area, commenting on the extension of the art into paintings on sheets of bark.

More than a century after the first European sighting, the enormous wealth and complexity of the rock art was finally recognised when an American-Australian Expedition into Arnhem Land studied the art on the plateau in 1948.

Since then a number of expeditions have examined the rock galleries of Kakadu, but it was some time before the paintings were seen in perspective.

George Chaloupka, of the Northern Territory Museums of Arts and Sciences, realised that 'aesthetic form, accompanied occasionally by an old myth, rather than the archaeological assessment of the art has preoccupied the majority of people working on Australian rock art in the past.

'The investigators of the Arnhem Land rock paintings interpreted and classified the presented body of art according to information given by their Aboriginal guides. The guides identified only the most recent horizon of paintings as being done by people; the earlier layers were said to have been painted by the spirits,' according to Chaloupka.

Confronted by more than a thousand surveyed sites, Chaloupka decided to consider the art from an art historian's perspective. His sophisticated classification divides Arnhem Land plateau rock art into four main periods: pre-estuarine, estuarine, freshwater and contact.

## PRE-ESTUARINE ART

This period is tentatively classified as beginning at around the height of the most recent ice age – about 20 000 years ago – though paintings that might represent extinct Australian fauna suggest it could be even older. The sea level rose after the glacial peak, developing the estuarine environment along the northwestern margin of the plateau and marking the end of this period of art.

The first images to appear on the rocks were grouped hand prints, some arranged into complex compositions, others high up on wall surfaces. Hand prints were made by placing the hand into wet pigment and pressing it on to the rock. By immersing grass and striking the rock surface with it, the Aborigines developed a new technique that may have arisen when grass seeds were a major ingredient of the local diet. This form was extended to the use of string or compressed paperbark. The string, wetted in red pigment, was thrown against the wall, leaving curvilinear images. Chaloupka suggested that imprints of more solid objects were in fact made by paperbark that had been compressed and impregnated before being thrown at the surface.

The earliest drawings in the Kakadu region are naturalistic images of animals and other subjects. Some of these examples are the first meaningful compositions in the region, revealing a purpose in patterns not seen in the hand and grass prints. In most of these drawings animals are sketched in an outline and coloured or textured in. They are often caricatures, with exaggerated features.

These early cartoonists concentrated on macropods, frequently painting the antilopine kangaroo (*Macropus antilopinus*), black wallaroo (*Macropus bernardus*), euro (*Macropus robustus*) and agile wallaby (*Macropus agilis*). Bushy-tailed and red-shouldered rock wallabies are also featured. Other animals such as the emu, the rock ringtail possum, the bandicoot, rock python, echidna and freshwater crocodile were also painted.

Images of the thylacine, the Tasmanian Devil and the Numbat (*Myrmecobius fasciatus*) have been found in this style, depicting animals that have been given mythological status by the Gagudju. The thylacine, for example, is said to have been associated with the Rainbow Snake, who was always accompanied by a pair of the 'dog bilong rainbow'.

Snakes and fish were seldom painted during this period. The snake appears only in a ceremonial context and only one small freshwater fish is shown. Other animals painted during the pre-estuarine period have been tentatively identified as extinct species. Of these, the most distinct portrait is of the long-beaked echidna (*Zaglossus*), which has been extinct in Australia for about 18 000 years.

The first human figures were also painted during this pre-estuarine period. Human figures took two forms. In one the figures are large, and detailed profiles are given to heads, feet and hands. Like animals, they are drawn with outlines and filled in with colouring such as ochre wash. In the other style the human male appears, according to Chaloupka, as a 'small, one-line thick stick figure in the act of spearing a large naturalistic animal'.

The stick-figure human form  has been termed *Mimi* art, as paintings of this kind were said to have been painted by spirit people called Mimi. The Mimi style is characterised by use of a single colour (generally red ochre) and frequent portrayal of groups of people, often fighting, hunting or engaged in rituals.

An unhafted, multi-barbed spear in two styles was used by these stick figures to hunt macropods and the freshwater crocodile. Another style of painting during this period was of 'dynamic' figures; the name stems from the expressive movement shown in the figures of males with spread legs and bodies thrust forward, expressing the physical motion of the hunter.

These are small, exquisite sketches of humans and animals and can be seen in more than 260 individual shelters across the plateau and outliers. Their fine detail and complex, story-telling compositions provide information about the social, economic and cultural lives of the inhabitants.

The women of this period carried long digging sticks and dilly bags, and occasionally spears and fire sticks. Men, depicted with unrealistically proportioned bodies, wore possum-hair belts much of the time and carried spears, boomerangs, clubs, stone axes and sticks. The spears were by then made of a single piece of wood, and the actual boomerangs used appear as stencils in many shelters. Stencils of spears, necklaces and dilly bags were also made.

The animal-headed beings painted during this period are the first evidence of mythogenesis in Kakadu rock art. These anthropomorphs, often resembling flying foxes, are usually depicted as participating with humans in various activities.

Animals were mainly macropods, but other species were also recorded including long-necked turtles and several birds and freshwater fish.

Another artistic development at this stage was the addition of dashes and marks emanating from the mouths of humans, animals and even some tracks. These depicted sensory experiences such as smell, force and sound.

Post-dynamic figures showed another sub-style, and were usually static silhouettes of humans. Later the human form became a one-line stick figure, though still wearing headdresses and skirts – demonstrating a continuation of the dynamic tradition together with the use of multi-barbed spears and boomerangs not used in earlier, larger naturalistic representations.

IAN MORRIS

Above: The X-ray style is well exemplified in these paintings of barramundi, an important food source for the Gagudju people.

Overleaf: Flooded swamplands were a valuable source of both animal and plant foods for Aboriginal people.

There are no distinctive drawings of yam plants at this stage, but at the end of the pre-estuarine period of art a style centred on several species of underground yam tubers began to appear. The figures show yam images transposed, stage by stage, into human and animal forms.

The Rainbow Snake, not portrayed in earlier art, became the second most often depicted subject; myths associated with this Dreamtime being are still integral to the ritual and legend of the region today.

The development of yam figures, including the Rainbow Snake, is almost certainly due to changes in the social and physical environment brought about by rising sea levels. The Rainbow Snake is associated in many myths with rain and floods, and emerges from the sea and eats or drowns people. Most species of yam require great amounts of rain for growth and may have re-colonised the Kakadu area with rising sea levels and increasing rainfall.

ESTUARINE ART

The drastic environmental changes that occurred 9000 to 7000 years ago heralded a new period of rock art.

The deep trenches of the freshwater creeks and rivers were flooded by the rising sea and filled with estuarine mud and clays, creating saltmarshes along the coast. The sea level rose slowly to reach – probably 6000 years ago – its present level. The Aborigines became largely dependent on rivers for food. One of their most important sources of food was the barramundi and this fish became the dominant subject of rock art (mullet, saltwater crocodile and lesser salmon catfish were also frequently depicted).

Wallaroos, small marsupials, and the emu abandoned the savannah and moved inland. The thylacine was no longer painted, and may already have become extinct or retreated inland.

As the environment and conditions changed, so did the tools of trade and hunting. Boomerangs were evidently no longer used, and hunters are portrayed using multi-pronged fish spears and multi-barbed halted and stone-headed spears. The spear thrower was introduced.

The art of this period also frequently depicts the black flying fox as an animal, not as a mythological being: a change in artistic perception probably inspired simply by hunting.

Another new character, Namargon – the Lightning Man – appeared at this stage: probably when the region first experienced enormous Wet seasons punctuated by violent electrical storms. Namargon was responsible for thunder, lightning and associated storms. When angry he would strike the ground with the stone axes that grew from his head, arms and knee joints, shattering trees, frightening the tiny Mimi spirits and sometimes killing people.

The distinctive and beautiful x-ray form of art began to assume dominance during this period, many thousands of years after the first simple forms had appeared. The x-ray style had two phases, descriptive and decorative, showing respectively an appreciation for intellectual realism and artistic talent.

### DESCRIPTIVE X-RAY

These highly detailed drawings reveal a detailed knowledge of anatomy: internal organs and bone structures are depicted along with the profile of the animal, and the human body has a skeletal form. The x-ray form even extended to inanimate objects from this period to the historical contact period when, for example, a rifle would be drawn with a bullet visible in its breech, or a Macassan kris outlined inside its sheath.

Again, the development of a style tells much about the lifestyles of the inhabitants at any one time. A stone-headed spear appears for the first time in descriptive x-ray style, and may therefore be associated with a point about 4100 years ago when the stone point appeared in the stone tool sequence.

The x-ray style developed in detail and sophistication, with sections and organs coloured differently and their interiors hatched. Mixtures of colour well beyond the basic ingredients were often used, and the introduction of white pigment showed a marked difference from earlier styles. Subjects were painted in white in full profile and internal body divisions and organs were then painted over with other pigments. In earlier styles outlines were first painted, usually in red, and lighter pigments such as yellow or white were only used later.

### THE FRESHWATER PERIOD

The most recent major environmental change in the Kakadu region took place about 1500 years ago (perhaps even more recently). On the subcoastal plains, the lower Magela and Nourlangie-Jim Jim creek systems developed freshwater billabongs and paperbark swamps. The disappearance of the saline plains meant

Rocky outcrops such as this provided shelter and security for the Aboriginal people of Kakadu.

that the creeks and waterholes became contiguous with tidal river reaches only during the Wet season floods.

The salt-marsh animals were no longer readily available, and the environmental change is found not only in shell middens, where freshwater mussels overlay saline creatures, but also in rock art.

As always, the economically important birds, animals and plants became the primary depictions of the art. In this period the magpie goose and the wandering whistling duck became important waterfowl species on the annually flooded wetlands. Representations of the magpie geese, long-necked turtles, file snakes and waterlilies joined the popular representations of barramundi and large catfish that had been introduced earlier into the gallery.

This period is also typified by the depiction of hunter's tools. The hunter is shown carrying a bundle of short 'goose spears' (a specialised weapon consisting of a short, light reed shaft tipped with hardwood), throwing sticks and a goosewing fan. Paintings also begin to portray rafts made from paperbark and used by women to collect goose eggs and waterlilies.

Most of the painting in this period is in x-ray style, but there are some in other forms such as the thick, one-line stick figure of a running hunter.

CONTACT

The contact period began with the visits of Macassan fishermen to the northern coast. These visits had a telling effect on the Alligator River and Arnhem Land Aborigines. The boats are depicted frequently in shelters close to the northern coastline, and the influence of contact is revealed further inland where Macassan possessions such as steel axes and knives were stencilled and painted.

European settlement brought, naturally, an enormous change in subject matter, though not in style: indeed, the only innovation was the infrequent use in very recent paintings of blue laundry dye.

The first images revealing the arrival of Europeans and their subsequent occupation of the region are of boats and exotic animals. Early explorers may have taken scant notice of the rock art they found but the inhabitants recorded the explorers, the first railway lines, buffalo hunters chasing their prey and the first missionary to enter the region. The buffalo industry can be blamed for the eventual disappearance of rock painting. The industry, which began in the

1890s, relied on a huge local labour force and only a few artists painted occasionally once Aboriginal culture began to erode.

Before that occurred internally painted and decorated hand stencils and beeswax designs were completed during the contact period. Many sorcery paintings, found mainly in the Oenpelli area, were also associated with this period. Chaloupka suggests these paintings are a direct result of introduced sicknesses: certainly, epidemics of measles, influenza and leprosy affected a large percentage of the population.

The x-ray style of painting faded, along with other forms of expression on the rocks, with the last x-ray paintings by a Maiali man, Barramundi Charlie, at two sites at Bourlangie in 1964.

Many young Aborigines have lost contact with their lands and traditional skills. There have been casual attempts using local clays and colours, but the alternative art form of bark painting, itself a long tradition, was taken up almost exclusively by the Aborigines who moved to live on missions or in settlements. Bark painting is now the dominant art form, though the enormous collection of early creativity in the form of rock art is still the region's major heritage.

Overleaf: Much of the Aboriginal artistic heritage is to be found in rock shelters and caves in the escarpment.

# THE LAND TODAY

## EUROPEAN SETTLEMENT AND EXPLOITATION

Above: Exotic animals have caused severe
damage to the park's fragile environment.

Left: Birds fly home to roost amid the
billowing monsoonal clouds.

The isolated existence of the Aboriginal people of Kakadu continued for tens of thousands of years in a relationship that was designed for sustainability and healthy survival.

But all that changed around the seventeenth century AD. From Celebes and Macassar to the North came Macassan fishermen in search of *bêche-de-mer* or trepang, the sea-slug considered a delicacy by the Chinese and a lucrative harvest for the fishermen.

The Macassans had a profound effect on the lives of the Aborigines, at first employing them and paying them with glass, rice, iron and tobacco. Smoking pipes and dug-out canoes were introduced by fishermen, and their word for canoe, *lepa-lepa*, became *lippa-lippa* in Arnhem Land. Other words used by the Macassans were adopted, including compass directions, tools and parts of boats. Aborigines often adopted Macassan names as well as their own.

Steel for making knives, spear blades and tomahawks was also brought by the visitors, who became part of the trading pattern of Arnhem Land. The fishing camps were often situated in defensive positions, and historical records from Indonesia testify to hostilities between Aborigines and the crews of the fishing praus; but there were also instances of Aborigines travelling to Macassar and returning after lengthy stays.

Dutch explorers also sailed along the north coast in the seventeenth century, sighting Aborigines and terming them 'savage, cruel, black barbarians who slew some of our sailors' (Willem Jansz, skipper of the *Duyfken*, 1606) and 'the most wretched and poorest creatures that I have ever seen in my age or time' (Jan Carstenz, discoverer of 'Arnhem's land' in 1623).

In 1803, Matthew Flinders charted the waters north of Australia, and sighted six wooden praus. He learned from the leader of the fleet that sixty praus and more than a thousand men were working on the Arnhem Land coast that season. Each prau collected about ten tonnes of trepang, and pearlshell, turtleshell, sandalwood and cypress wood were also gathered.

The Macassans, who travelled to Australia on the northwest monsoon winds, returned home every year on the southeast trades, leaving no permanent camps or factories. But they left indelible impressions and some tangible legacies, such as the tamarind tree.

The swollen waters of a permanent billabong spill over on to the surrounding floodplain.

Above: In elegant and graceful formation, brolgas fly through the humid
Wet season air.

Overleaf: The flooded East Alligator River forces its way through the
escarpment channel to the floodplains beyond.

The next stage of outside influence started in the 1820s, when the British built Ford Dundas on Melville Island in 1824. The fort was abandoned in 1829 – and together with tonnes of useless stores, the departing garrison left an unspecified number of Timorese buffalo to fend for themselves. The buffalo throve in the climate that had so adversely affected European settlers.

Fort Wellington, at Raffles Bay on the Cobourg Peninsula, lasted from 1827 to 1829 and Port Essington, west of Raffles Bay was established in 1838 and became the North's largest settlement before its demise in 1849.

Captain Phillip King had surveyed the northern coastline of Australia in 1818 and on May 6 that year he entered the East Alligator. 'The mangrove bushes on the banks of the river, which was quite salt, were crowded with the nests of an egret in which the young birds were nearly fledged,' he wrote. 'Hawks, wild ducks, pelicans, and pigeons were also abundant, and an immense flight of white cockatoos hovered over the mangroves, and quite disturbed the air with their hideous screamings.

'We encountered several very large alligators [saltwater crocodiles] and some were noticed sleeping on the mud … No inhabitants were seen, but the fires that were burning in all directions proved that they could not be far off.'

He then journeyed to what must be the South Alligator River and noted that, '… Alligators were as numerous as in the other river, whence the name of Alligator Rivers were bestowed upon them.'

King noted the 'great tides', and that the land was no more than three feet (90 centimetres) above the high water mark, apart from a few 'wooded hills' and some groups of trees including palm trees. The landscape was monotonous, and the view 'to say the best of it, unvaried and heavy'.

King explored the Alligator Rivers but the hinterland was not studied thoroughly until Ludwig Leichhardt, journeying from Moreton Bay to Port Essington (a distance of 5000 kilometres) entered the region late in 1845. Leichhardt waxed lyrical about the region's wildlife and described many new plants and animals. Some, such as species of *Grevillea* and *Melaleuca* and a grasshopper that was named after him, have only recently been rediscovered.

He described the South Alligator valley as 'a beautiful valley which lay before us like a promised land' and as 'magnificently grassed'.

Leichhardt made contact often with Aborigines, who guided him to water wells, edible fruits, and across swamps. 'The natives were very numerous', he wrote, 'and employing themselves either in fishing or burning the grass on the plains, or digging for roots.'

After crossing the East Alligator River Leichhardt met some of the region's feral buffalo. 'These are the offspring of the stock which had either strayed from the settlement at Raffles Bay, or had been left behind when that establishment was broken up,' he wrote. As he headed toward Port Essington, he saw buffalo emerging three and four at a time from waterholes.

Gold was mined at Yam Creek and silver at Pine Creek in 1885, with finds of tin, lead and antimony attracting a few prospectors and miners.

Buffalo hunting began in the late 1890s in the Alligator Rivers region, with the colourful figure of Paddy Cahill leading the way and taking up residence at Oen-pelli. Each operator was supported by a labour force of a hundred or more Aborigines – the first major departure from the traditional way of life for the Aborigines of Kakadu.

Cahill, a renowned buffalo hide trader, established his property in 1906; the Commonwealth purchased it in 1916 before abandoning it three years later. In 1920, the Arnhem Land Aboriginal reserve was declared in the Oenpelli area.

In 1912, Cahill welcomed Professor Baldwin Spencer, who took hundreds of photographs at the Oenpelli lagoon and wrote of barramundi fishing and the catching of water-snakes. He was particularly impressed with the bark painting of the Kakadu people and commissioned, with sticks of tobacco as payment, fifty works of art by the best local artists.

Spencer also met with Geimbio, Kulunglutchi and Umoriu people, and noted the custom of men having up to seven wives. Baldwin's diaries illustrate the relationship developing between the Aborigines and Europeans: the Aboriginal men became aware of European concern about the number of their wives, began to apportion the women out and to hide some of their wives.

Since the days of Paddy Cahill and Baldwin Spencer a steady stream of fossickers, explorers, buffalo and crocodile shooters, cattlemen, timber cutters, safari operators, miners, missionaries, fishermen, scientists, government officials and tourists have entered the Kakadu region. The most significant recent

Paperbarks stand like sentinels in the flooded swampland, where the water ebbs and flows with the monsoonal rains.

influences on the area have been the granting of title in 1978 of nearly the entire area of Kakadu National Park to the Kakadu Aboriginal Land Trust and the subsequent proclamation, on April 5, 1979, of Stage One of Kakadu National Park.

LAND DEGRADATION

The wilderness and wildlife qualities of Kakadu have remained intact in large areas despite the introduction of new species and new management practices, but the encroachment of contemporary activities is increasing the strain by removing elements and adding alien factors such as buffalo, which have greatly altered the region's delicate ecological balance.

The effects of human activity, however, go beyond what park rangers can control. Projects such as the Ranger uranium mine, for example, have to be dealt with in a political and bureaucratic atmosphere that covers a far greater area than that controlled by the National Parks and Wildlife Service.

One of the most obviously adverse ecological effects of human presence in the area is the introduction of exotic animals, most notably buffalo, pigs, cattle, brumbies, cats, dogs and donkeys.

Buffalo adapted all too well to their new habitat, and the only threat to their numbers came from human hunters. A hide and horn industry began in the 1880s, based at Oenpelli. Paddy Cahill, the major buffalo hunter of the region, introduced shooting from horseback. Such was the development of the industry that the *Northern Territory Times* reported in 1898 that, 'as a source of profit, buffalo shooting is said to be on its last legs, as this year should about finish the inglorious slaughter'.

Wishful thinking. The buffalo population continued to flourish, along with the industry. By 1911 more than 100 000 hides had been exported; between 1911 and 1956 the figure rose to an average of 7000 hides a year. In 1937-8 alone, 16 549 hides were exported in twelve months.

By 1956 the trade in hides had become uneconomical, and the industry diverted to meat production for humans and pets. The buffalo had by then become an integral part of Territory folklore – but it was also a large-scale destroyer of the very environment it symbolised.

About 200 000 buffalo roamed the Territory in the 1960s. By the early 1980s their numbers were estimated to have increased by another 50 000.

Buffaloes have wallowed below this giant fig, creating murky pools that threaten the root systems of this ancient tree.

Introduced animals threaten the fragile ecology of Kakadu.
Feral pigs (top) and buffaloes (right) trample and destroy
vegetation, and introduce weeds and diseases; brumbies
(above) are at present a more limited problem.

Buffalo are truly bulls in an environmental china shop. Their wallowing, trampling, and grazing have damaged and changed vegetation patterns, turning large areas into mud wallows. On tidal flats and floodplains, wallowing and trail-blazing breach the natural levees of the river and creek systems, leading to excess flooding and ponding – which in turn damages or destroys freshwater communities. Insect numbers increase in the brackish water, and paperbark swamps have been wiped out in some areas by a buffalo-induced rise in salinity.

By eating out the more desirable, high-protein grasses such as *Hymenachne*, buffalo promote the growth of less palatable plants. Native herbivores that depend on high-protein forage suffer more than the buffalo: only since buffalo culling began have these high-protein grasses begun to re-introduce themselves on the floodplains, leading to the return of magpie geese and other birds.

Many birds have been affected by the presence of buffalo. As well as the magpie goose, the scrubfowl, rainbow pitta and yellow chat have all decreased in numbers and CSIRO research shows a majority of species have suffered a reduction in density due to habitat damage by buffalo.

In Stage Two of the park, buffalo exclusion plots installed by the CSIRO at Kapalga have led to a marked regeneration of *Hymenachne* on the edge of plains from which it had almost disappeared. The ground cover regenerated from runners that had not been accessible to buffalo and that could now flourish in the less compacted soils.

These experimental areas have also shown a dramatic increase in paperbark, eucalypt and pandanus understoreys. In other areas native plants such as the red lotus lily have regenerated.

Buffalo trails, grazing, trampling and wallowing cause erosion by destroying vegetation cover and without the protection of plants, the fragile soils of Kakadu are washed into creeks and rivers. Since large-scale buffalo culling began, the Wetlands are rapidly returning to their natural state.

As if its more obvious environmental effects were not enough, the buffalo may also encourage the spread of exotic and noxious weeds and diseases. By disturbing the soil and distributing seeds in their hair and excreta, buffalo help spread noxious weeds such as *Hyptis suaveolens* and the pasture legume, Townsville stylo (*Stylosanthes humilis*). Buffalo are potential spreaders of bovine diseases

Drenched by moisture, the glowing colours of this eucalypt shine amid the surrounding greenery.

such as brucellosis, lice, kidneyworm, tapeworm, cattle ticks, buffalo fly and anthrax, making efficient control of these diseases difficult and threatening the beef industry in northern Australia.

The feral pig is creating similar problems in the Kakadu region by wallowing and rooting. Moreover pigs consume native fauna – including small mammals, frogs, snakes and birds' eggs – and plantlife such as tubers, shoots and fruits.

Pigs also share the buffalo's trait of carrying disease and the similarity in habits between the two animals has led to fear that the reduction in buffalo numbers may lead to an increase in the pig population.

A far greater problem is the spread of exotic weeds. About seventy species have been recorded in Kakadu, and several now present a major problem.

The thorny central American *Mimosa* is thought to have been introduced into the Northern Territory about 1940, and now covers about 8000 hectares, half of that in one single infestation near the Adelaide River. There are also infestations inside Kakadu National Park; *Mimosa* grows rapidly, produces large numbers of long-lived, easily transported seeds and is highly tolerant of drought and flood.

The other weed causing great concern in the area is *Salvinia molesta*, an aquatic fern from Brazil. This pernicious plant threatens tropical freshwater habitats around the world, and the National Parks and Wildlife Service is experimenting with biological controls in an attempt to eliminate it.

But the introduced weed spreads quickly, blocking waterways and preventing supply of nutrients to native plants. Harvesting and herbicides have proved unsuccessful, so weevils from Brazil have been brought in to control it. Similar methods have been used to control another aquatic weed, the water hyacinth (*Eichhornia crassipes*), which is not – yet – a major problem in Kakadu.

Although the Aboriginal pattern of regular burning may have led to a decrease in the amount of rainforest and an expansion of woodland and open forest, modern rates of denudation in the region are relatively low. Researchers say closer settlement may not cause a drastic increase in erosion if there is minimal destruction of plant cover, but if extensive areas of soil are exposed as a result of mining, road-building and construction erosion will increase by between 20 and 100 per cent in the disturbed areas. Obviously an increase of such magnitude will have long-term effects on Kakadu's already threatened ecological balance.

Human presence in the region has been an undeniable influence on the environment, and perhaps the most obvious influence has been that of mining. Several areas are excised from the park for mining, including the operating Ranger uranium mine. The proposed Jabiluka mine is also on the boundaries of, and surrounded by, the park; like Ranger it is in the Magela Creek catchment, which enters the East Alligator River. The other proposed mine, Koongarra, is in the South Alligator River catchment.

The physical impact of the Ranger mine, a giant scraping across a sensitive tract of land, is obvious, but the impact of the mine on areas outside the site is also a threat to the ecology of the area. A major concern is the potential leakage or release of contaminants into the Magela Creek system. Ranger has a 'restricted release zone', from which no materials, including water, are supposedly released into the system. But the release of water is an essential part of the mine's operations and by 1987 the problem of accumulation was extreme. Rainfall had been higher than expected and evaporation more limited, thus creating storage problems in the retention ponds. A study commissioned by a government–industry working group failed to come up with a solution that would totally contain the water on the mine site.

This dilemma is symptomatic of the problems associated with the existence of a large mine in a sensitive environment. The mine is a large unbalanced element in an otherwise balanced environment. However, it exists and all concerned with the future of Kakadu must now work toward minimising the adverse ecological and social impact it has on the Kakadu National Park and its Aboriginal owners.

Overleaf: The rains have stopped, the sun bursts forth;
the plants respond with dynamic growth as they compete
for nutrients in the endless tropical life-cycle.

Above: Glinting in the sun, water seeps from the rocks for
several days after rain has finished.

Left: A tree bursts into flower at the top of the escarpment.

The treetops of the monsoonal rainforest in the foreground give way to
isolated rocky outcrops and thence to the endless flat floodplains that
stretch to the coast.

Above: New grasses shoot from the swamp as cloud sweeps in from
the coast.

Overleaf: The diversity of Kakadu's landforms contributes to its unique
value as a heritage area.

# A PARK FOR THE FUTURE

CONSERVING A NATURAL WONDER

Above: Grevillea blooms in the Wet season
at the rocky foot of the escarpment.

Left: Jim Jim Falls, whose waters fluctuate
with the rhythm of the seasons.

The fine balance of nature determines that a wetland area or an expanse of rainforest in a river catchment are only as protected as the upper regions of that catchment. A river is only as pristine as its source.

Accordingly, the protection of the upper catchment and headwaters of the South Alligator River is an important element in establishing a fully protected regional National Park, especially when that southern region contains cultural and natural features that are represented poorly or not at all in Stages One and Two of Kakadu.

The originally proposed Stage Three of Kakadu National Park, consisting of all of Goodparla pastoral lease and part of the neighbouring Gimbat lease, is an environmentally strategic area at the headwaters of the South Alligator River, to the south of the first two stages of the National Park.

The Federal government decision in 1987 to exclude thirty-five per cent of the proposed third stage for the purposes of mining exploration means a significant part of the South Alligator's upper catchment will now be outside the park but included in a conservation zone. Environmental disturbances in that area will still affect the park itself.

The area originally proposed as Stage Three contains sacred sites, rare and endangered species, unusual land forms and spectacular escarpment country. Moreover, its position at the headwaters of the enormous South Alligator River system means that any disturbance will flow on literally and figuratively to the lowlands, floodplains, coastal swamps and tidal flats in the World Heritage Area to the north. The wetlands of Stages One and Two would be affected directly. The close ecological links within a river catchment have long been evident in the South Alligator River valley: high sediment loads in water entering Stages One and Two during the wet season from Goodparla Station are largely due to erosion at Goodparla through overstocking and buffalo damage.

The originally proposed Stage Three, comprising about 6700 square kilometres, is the only adequate representation of the southern hills and basins topographies in the Alligator Rivers region. Nine land systems, representing 1060 square kilometres, are not found in Stages One and Two, and a further five are poorly represented. Those five comprise a further 2975 square kilometres of Gimbat and Goodparla: in other words, sixty per cent of the leases contains land

Water swirls and eddies in a whirlpool as it is channelled
from the escarpment to the flatlands below.

systems absent in whole or part from Kakadu National Park Stages One and Two.

That sixty per cent comprises areas of tall open forest, woodlands, mixed scrub, sandstone monsoon forests, mixed open forest with monsoon elements, spinifex and fringing forests. National Parks and Wildlife Service (ANPWS) research indicates that these diverse plant communities, and the spatial and topographic complexity of the Gimbat and Goodparla area, are vital to the regional conservation of animal communities.

While tracts of woodlands and open forest such as those found in Gimbat and Goodparla may be less spectacular than escarpments, wetlands or rainforests, they are nevertheless important habitats for animals and plants. Fauna surveys have determined that they are more important to the maintenance of plant diversity than other Kakadu forests.

The animals of Gimbat and Goodparla are also significant. The first two stages of the park are considered vital for the conservation of waterfowl, crocodiles, fishes and mammals: but some of these species and their environments are conserved only marginally in those stages. The Australian National Parks and Wildlife Service contends that such species, which either have limited distributions or are uncommon or rare in the region, would be more adequately conserved with the inclusion of Gimbat and Goodparla.

At least one animal may be unique to the Gimbat region. A new and as yet undescribed species of native mouse (*Pseudomys* sp.) discovered in Gimbat has not been recorded elsewhere.

Other mammals would also be more protected by the inclusion of the entire proposed Stage Three. The black wallaroo, endemic to the sandstone escarpment and outlying massifs of western Arnhem Land, is found in a small section of the escarpment in Stage One and an area of sandstone rock country in Gimbat, making inclusion of that second area of undoubted assistance to its conservation. The Nabarlek wallaby, the rock possum, the large rock rat and Harney's marsupial mouse all have very small distributions and their survival would be improved by National Park protection.

The ANPWS considers much of the area earmarked for mineral exploration as having major cultural significance, primarily because of the presence of a large number of Aboriginal sacred sites and rock paintings. One group of sacred sites

In the late afternoon sun, the foliage of the rainforest provides a striking contrast to the colours of the escarpment.

is a contentious area at Coronation Hill which, under the 1987 decision to exclude thirty-five per cent of the proposed Stage Three for mineral exploration, was not included in the national park extension.

The landscape of the region is considered by its Aboriginal inhabitants to be the work of the spiritual beings: Bula, one such creator, is said to have 'made' several sacred and/or dangerous sites during a journey through this area. If interfered with, these sites can cause massive destruction.

Sites associated with Bula should, the Aborigines say, remain undisturbed. Within the sites are areas with different levels of importance and 'danger'. There are many 'link-up' areas where mining would disturb Bula.

'Djang andjamun' sites are areas where the creator beings or other forces remain active, are dangerous and should only be visited by those who can perform appropriate rituals in the correct language.

There is also a rich array of Aboriginal art in the area. More than 100 sites have been found among the escarpment gorges from Waterfall Creek to Gimbat Creek: as this is the only area systematically surveyed many more sites are likely along the escarpment and elsewhere.

The total number of sites is considered to at least rival that of Stage One and may surpass Stage Two, but many, such as those at Christmas Creek, have been damaged by buffalo and by uncontrolled human visitors.

An important element of the cultural heritage of the Gimbat and Goodparla area is the Aboriginal distinction between that region and areas to the north. The Gimbat people are culturally distinct from the people to the north; a difference reflected in their art and archaeological sites, making these a valuable addition (not a superfluous extension) to sites protected in Stages One and Two.

X-ray paintings in Gimbat show a different subject range and a variety of artistic conventions which, in European art terms, would represent a distinct 'school'. There are abstract designs more typical of the art of the south. Some of the art in Stage Three appears to be older, dating from the Pleistocene period and extends the range of styles more common in northern areas.

The pressures of protecting and managing such a varied landscape and storehouse of wildlife are immense, and have been handled magnificently by the park's rangers and administrative personnel.

The Ranger mine, where production of uranium began,
amid considerable controversy, in 1979.

But the pressures continue. Miners, for example, argue that 'multiple land-use' should be allowed in the Park. There is nothing 'multiple' about mining as a land use, however. It destroys habitats; rehabilitation of mining areas, however efficient, cannot restore the natural ecological balance, and the disturbance created by mining extends in time and space well beyond the mining operations.

The cultural significance of Kakadu is in no way inferior to, or separate from, its natural qualities. The area houses Australia's greatest collection of Aboriginal artwork and forms an archaeological study area where significant discoveries about the past have already been gleaned. With those qualities are an enormous number of sacred sites. Many are relatively unknown to European Australians and may well only be revealed when threatened by mining or some other form of development. Many find it difficult to realise that the very nature of these sites makes their secrecy vital, prompting cries about opportunism when the sites are reluctantly disclosed under threat of disturbance or destruction.

The international significance of Kakadu can be measured by its inclusion on the World Heritage list, by the World Heritage Committee's encouragement for more of it to be placed on the list, and by the inclusion of Kakadu's wetlands on the United Nations list of Wetlands of International Importance especially as Waterfowl Habitat.

Such international recognition notwithstanding, we must learn to appreciate the true significance of protecting areas such as Kakadu. The International Union for the Conservation of Nature's 1980 World Conservation Strategy found that 'tropical dry or deciduous forests' are among areas with 'concentrations of threatened species'. Similarly, 'tropical grassland savannah' is one of six ecosystems 'unrepresented, or poorly represented' in protected areas. These considerations serve to further increase the demand for good management of Kakadu, even without taking into consideration its many other qualities.

The maximisation of the diversity of native species and communities is a proper objective for park management and, coupled with the protection of landscapes, cultural qualities and catchment areas, is a worthy aim. Such a management philosophy will, one can hope, maintain Kakadu as a unique part of Earth, offering a dazzling and unparalleled array of Australia's cultural heritage, environments, landscapes, animals and plants.

Above: Mt Brockman, an Aboriginal site of enormous spiritual significance and gallery of fine rock paintings, lies just a short distance from the Ranger mine.

Overleaf: The tree-flanked branches of the East Alligator River wind slowly to the coast, their waters becoming more brackish with the tidal ebb and flow.

Above: Cooinda, a tourist centre at Yellow Water, near the conjunction of
Jim Jim Creek and the South Alligator.

Left: The sensitive development of tourist activities and facilities may be
an important step in gaining support for the preservation of Kakadu.

Overleaf: Approaching thunderclouds threaten to bring more rain to an
already flooded Jim Jim Creek.

Above: Creeping exposed roots explore the rockface in
search of nourishment.

Right: Cascading waters flow across the escarpment, creating
natural patterns of subtle beauty.

Above: On the edge of the monsoonal rainforest, plants compete for light, creating an infinite variety of forms and patterns.

Left: The intricate growth of this powerful fig is a response to its endless search for light and nourishment.

Overleaf: Weathered by eons of sun, rain and wind, the edge of the escarpment has broken down and its sharp contours have worn away.

Above: Acacia, generous in both colour and nectar, springs from barren
rock and contributes to the ecological balance of Kakadu.

Left: Nature prepares a rainforest garden whose beauty and variety are
unmatched by any creation of human beings.

Overleaf: An instant ago it was raining; now the sun beats down and
steam rises from the rock to form a mist over the lowland hills.

The weathering of these timeless rocks has created cracks
and caves and deep gorges where small creatures of many
kinds make their homes.

Above: The vigorous fresh growth of swamp grasses provides
a perfect environment for nesting waterbirds.

Overleaf: As the sun sets over the Magela swamp, waterbirds
gather and squabble before roosting for the night.

Above: The still beauty of the Nourlangie swamp is broken only by the
first drops of a tropical rainstorm.

Left: With primeval intensity, a monsoonal storm sweeps over the
coastal floodplains.

Overleaf: A row of dense mangroves marks the conjunction of the
Kakadu floodplains and the coastal estuary, where the tidal waters of the
Arafura Sea meet the northern edge of Kakadu.

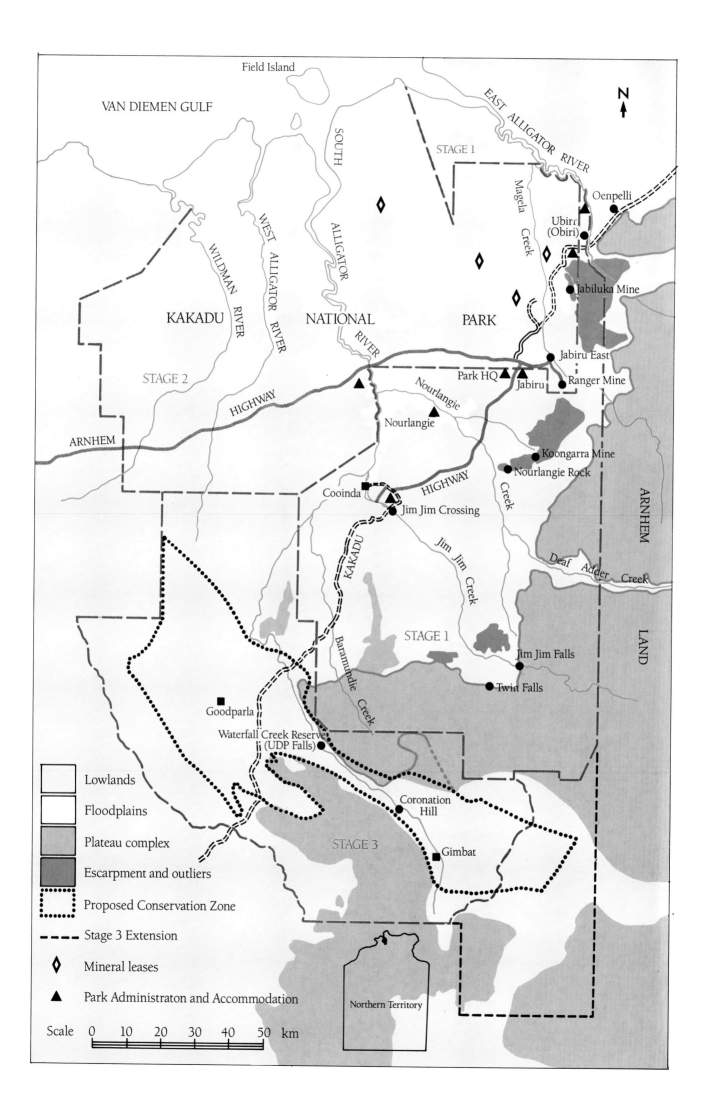

Field Island

VAN DIEMEN GULF

STAGE 1

EAST ALLIGATOR RIVER

N

Oenpelli

Ubirr
(Obiri)

Jabiluka Mine

WILDMAN RIVER

WEST ALLIGATOR RIVER

SOUTH ALLIGATOR RIVER

Magela Creek

KAKADU

NATIONAL

PARK

STAGE 2

Jabiru East

Park HQ

Jabiru

Ranger Mine

ARNHEM

HIGHWAY

Nourlangie

Nourlangie

Koongarra Mine

Nourlangie Rock

ARNHEM

HIGHWAY

Cooinda

Jim Jim Crossing

Creek

Jim Jim Creek

KAKADU

Baramundie Creek

Deaf Adder Creek

LAND

STAGE 1

Jim Jim Falls

Twin Falls

Goodparla

Waterfall Creek Reserve
(UDP Falls)

Coronation
Hill

STAGE 3

Gimbat

Lowlands

Floodplains

Plateau complex

Escarpment and outliers

Proposed Conservation Zone

Stage 3 Extension

Mineral leases

Park Administraton and Accommodation

Northern Territory

Scale    0    10    20    30    40    50    km

# ACKNOWLEDGMENTS

The task of gathering information on such a culturally and ecologically diverse part of our country was made easier by the generous help of many people. Grateful recognition and thanks go to many in the Australian National Parks and Wildlife Service (ANPWS), especially Dan Gillespie, assistant director of Northern Operations. Dr Tony Press, Gordon Anderson and Des Pike of ANPWS also contributed their comments, and information obtained from them and other staff and research papers on a variety of subjects reaffirmed to the author the vital national role of ANPWS.

I also wish to thank Robert Story, who permitted me to cite his work for ANPWS. The information and guidance obtained from *Kakadu: A World Heritage of Unsurpassed Beauty*, by Professor Derrick Ovington, director of ANPWS, is gratefully acknowledged. I would like to thank Dick Braithwaite of CSIRO for providing many papers on Kakadu and its ecology. Peter Alston and Michelle Albrecht also found valuable research material.David Cooper of the Aboriginal Sacred Sites Protection Authority in Darwin provided information on the sensitive matter of sacred sites. The work of George Chaloupka, curator of Aboriginal rock art, Museum and Art Galleries of the Northern Territory, is also gratefully acknowledged.

I would also like to thank Chris Harris, Yvonne Forrest, Lyn Allen, Janelle Lassig and Melanie Thiedeman for their help and information.

At the Australian Conservation Foundation, Carole Owen and Prue Lamont were a great help in the library and the director, Phillip Toyne, provided guidance.

A very special thank you to Fran Coughlan, who took on the enormous role of research assistant and achieved an enormous amount. Without her, the book would not have been completed. Her canine offsider, Tippa, provided light relief when it was most needed. SIMON BALDERSTONE

My task as photographer in Kakadu was made easier by the willing assistance of many people. I would like to thank the director, ANPWS, for permission to undertake this work in Kakadu National Park. Thanks go also to the rangers and staff of Kakadu and the Gagadju Association and to the staff at CSIRO Tropical Ecosystems Research Centre, Kapalga field station.

Special thanks are due to the following people for their help and cooperation in the photography of this book: Dr Tony Press, Ian Morris, Greg Miles, Jane Moore, John A. Estbergs, Rolf Gerig, Peter Brady, Gus Wanganeen, Peter Panguee, management and staff Four Seasons Cooinda Hotel/Motel resort, Phillip Burt, Don Christophersen, Kakadu Air Services, Boatland Winnellie, Simon Taylor, J. & M. Groves, Woolner Station helicopter hire, Gary Dew and Kerry Dibben for pulling me out of the mud, and Carlia and Bunitj for rearing and taming the phascogale. LEO MEIER

# INDEX

## PHOTOGRAPHIC NOTE

For the images in this book Leo Meier used Nikon F3 and FA cameras and the following Nikkor lenses: 18mm f3.5, 24mm f2, 28mm f2.8, 35mm f1.4, 50-300mm f4.5 ED, 55mm f2.8 micro, 85mm f1.4, 105mm f2.8 micro, 180mm f2.8 ED, 200mm f4 micro, 300mm f2.8 IF-ED, 600mm f4 IF-ED. All the equipment performed flawlessly under the most extreme conditions of heat, dust and the humidity of tropical monsoon weather.

Films used were Kodachrome 25 PKM and 64 PKR.